Essential
PREALGEBRA
Skills Practice Workbook

$$7x - 3 = 5x + 9$$

$$3x + 2 > 8$$

$$6 - 4 \cdot 3 + (2 + 3)^2$$

$$\frac{4}{11} = 0.\overline{36}$$

$$\frac{7}{8} = 0.875 = 87.5\%$$

$$\frac{15}{x} = \frac{5}{6}$$

Chris McMullen, Ph.D.

Essential Prealgebra Skills Practice Workbook

Chris McMullen, Ph.D.

Zishka Publishing
ISBN: 978-1-941691-08-3

Mathematics > Prealgebra
Study Guides > Workbooks > Math

CONTENTS

INTRODUCTION

The goal of this workbook is to help prepare students for a successful transition from arithmetic to algebra. Following is a sample of what to expect:

- Improve arithmetic fluency. Since arithmetic is used to solve algebra problems, it is important for students to be fluent in their arithmetic skills. The exercises in this workbook involve a variety of arithmetic skills to help build fluency.

- Fractions, decimals, and percents. Fractions come about naturally when doing algebra. This workbook has separate chapters devoted to fractions, decimals, and percents.

- Build a strong foundation. Fundamental algebra concepts – such as combining like terms, isolating the unknown, and factoring – are introduced. Developing a solid foundation is essential toward mastering the subject of algebra.

- Beginning algebra practice. Chapters 6-7 focus on simplifying basic expressions and solving simple equations. This lets students learn to swim in the shallow end before diving into the deep end.

- Thinking through math. Students who succeed at higher levels of mathematics learn to think their way through the math. Strive to get your students to think through the exercises. Students who try to understand the ideas and the logic behind the examples are more likely to be successful.

- Answer key. Practice makes permanent, but not necessarily perfect. Check the answers at the back of the book and strive to learn from any mistakes. This will help to ensure that practice makes perfect.

1 EXPONENTS

An **exponent** appears to the top right of a number, like the 2 in 4^2 or the 5 in 3^5. An exponent is basically shorthand for repeated multiplication, as shown by the examples below.

$$4^2 = 4 \times 4$$
$$2^3 = 2 \times 2 \times 2$$
$$7^4 = 7 \times 7 \times 7 \times 7$$
$$3^5 = 3 \times 3 \times 3 \times 3 \times 3$$

As shown above, an exponent tells you how many times to multiply a number by itself. For example, 10^6 means to have 6 tens multiplying each other:

$$10^6 = 10 \times 10 \times 10 \times 10 \times 10 \times 10$$

A **power** is another term for exponent. In 8^3, we may call 3 the exponent or we may call it the power, whereas 8 is called the **base**.

When the power equals 2, we call it a **square**. For example, 6^2 reads as "6 squared."

When the power equals 3, we call it a **cube**. For example, 8^3 reads as "8 cubed."

The symbol $\sqrt{}$ is called the **radical sign**. It is used to indicate a root.

A **square root** asks, "Which number multiplied by itself equals the value under the radical?" For example, $\sqrt{25}$ asks, "What number times itself equals 25?" One answer is $\sqrt{25} = 5$ since $5 \times 5 = 25$.

A **cube root** asks, "Which number cubed equals the value under the radical?" We write a small 3 to the left of the radical sign to indicate a cube root. For example, $\sqrt[3]{64} = 4$ is the cube root of 64 since $4^3 = 4 \times 4 \times 4 = 64$. In contrast, $\sqrt{64} = 8$ is the square root of 64.

Other roots can be similarly indicated. For example, $\sqrt[4]{81}$ is the fourth root of 81. The answer to $\sqrt[4]{81}$ is 3 because $3^4 = 3 \times 3 \times 3 \times 3 = 81$.

A square root is basically the opposite of a square. For example, compare $4^2 = 4 \times 4 = 16$ to $\sqrt{16} = 4$. Similarly, a cube root is basically the opposite of a cube. For example, compare $2^3 = 2 \times 2 \times 2 = 8$ to $\sqrt[3]{8} = 2$.

- What happens if the power is 0 or 1, like 5^0 or 9^1? Good question. See Sec. 1.3.
- Do exponents really save that much time? You bet. Try writing 3^{100} the long way.
- Why does it say, "One answer is $\sqrt{25} = 5$"? Is that because there is another answer? Yes! See Sec. 1.6.
- Can you have a negative power, like 2^{-1}? Yes, you can. See Sec. 1.5.
- Can you have fractional powers, like $8^{2/3}$? Yes! See Chapter 3.

1.1 Squares

To square a number, multiply the number by itself.

Example 1. $7^2 = 7 \times 7 = 49$ **Example 2.** $13^2 = 13 \times 13 = 169$

1) $2^2 =$ 2) $5^2 =$

3) $8^2 =$ 4) $1^2 =$

5) $6^2 =$ 6) $3^2 =$

7) $9^2 =$ 8) $0^2 =$

9) $4^2 =$ 10) $10^2 =$

11) $15^2 =$ 12) $25^2 =$

13) $11^2 =$ 14) $14^2 =$

15) $30^2 =$ 16) $18^2 =$

17) $16^2 =$ 18) $12^2 =$

19) $20^2 =$ 20) $19^2 =$

1.2 Cubes

To cube a number, multiply the number by itself two times.

Example 1. $6^3 = 6 \times 6 \times 6 = 36 \times 6 = 216$
Example 2. $11^3 = 11 \times 11 \times 11 = 121 \times 11 = 1331$

1) $3^3 =$

2) $1^3 =$

3) $7^3 =$

4) $5^3 =$

5) $10^3 =$

6) $2^3 =$

7) $12^3 =$

8) $0^3 =$

9) $15^3 =$

10) $9^3 =$

11) $13^3 =$

12) $4^3 =$

13) $20^3 =$

14) $8^3 =$

1.3 Powers

The power tells you how many times to multiply a number by itself. Follow the examples.

Example 1. $5^4 = 5 \times 5 \times 5 \times 5 = 25 \times 25 = 625$

Example 2. $3^6 = 3 \times 3 \times 3 \times 3 \times 3 \times 3 = 9 \times 9 \times 9 = 81 \times 9 = 729$

Example 3. $2^5 = 2 \times 2 \times 2 \times 2 \times 2 = 4 \times 4 \times 2 = 16 \times 2 = 32$

What about a power of one? That's easy: $7^1 = 7$. Here, there is just one 7.

What about a power of zero? That's also easy, but trickier to understand: $10^0 = 1$. Why does $10^0 = 1$? Consider that $10^2 = 100$, $10^3 = 1000$, $10^4 = 10,000$, $10^5 = 100,000$, etc. Note that $10^m \times 10^n = 10^{m+n}$. (You can verify this rule for any values of m and n. Try it! For example, $10^2 \times 10^3 = 100 \times 1000 = 100,000 = 10^5$.) If you plug $m = 2$ and $n = 0$ into the formula, you'll get $10^2 \times 10^0 = 10^2$, which becomes $100 \times 10^0 = 100$. In order for $100 \times 10^0 = 100$ to be true, we need $10^0 = 1$. This is one way to see that raising any nonzero number to the power of zero is equal to one. For example, $3^0 = 1$.

1) $2^7 =$

2) $4^5 =$

3) $8^4 =$

4) $6^5 =$

5) $11^0 =$

6) $7^4 =$

7) $2^9 =$

8) $9^4 =$

9) $20^1 =$

10) $13^3 =$

11) $5^5 =$

12) $10^6 =$

13) $3^7 =$

14) $100^2 =$

1.4 Negative Numbers

A negative number times another negative number is positive. Therefore, the square of any negative number is positive.

A negative number cubed is negative. There are three minus signs multiplying one another (see Example 2 below). Two of these signs cancel, leaving one minus sign in the answer.

If a negative number is raised to an even power (like 2, 4, 6, etc.), the answer is positive.

If a negative number is raised to an odd power (like 1, 3, 5, etc.), the answer is negative.

Example 1. $(-8)^2 = -8 \times (-8) = 64$

Example 2. $(-4)^3 = -4 \times (-4) \times (-4) = 16 \times (-4) = -64$

Example 3. $(-10)^4 = -10 \times (-10) \times (-10) \times (-10) = 100 \times 100 = 10{,}000$

Example 4. $(-3)^5 = -3 \times (-3) \times (-3) \times (-3) \times (-3) = 9 \times 9 \times (-3) = 81 \times (-3) = -243$

1) $(-6)^3 =$

2) $(-5)^4 =$

3) $(-2)^6 =$

4) $(-7)^3 =$

5) $(-4)^6 =$

6) $(-10)^5 =$

7) $(-8)^4 =$

8) $(-11)^3 =$

9) $(-9)^4 =$

10) $(-12)^2 =$

11) $(-2)^7 =$

12) $(-3)^6 =$

13) $(-20)^3 =$

14) $(-5)^5 =$

1.5 Negative Powers

A power of -1 makes a reciprocal. The reciprocal of an integer is just one over the integer. For example, $5^{-1} = \frac{1}{5}$. (We'll learn more about reciprocals in Chapter 3. For now, all you need to know about them is that for an integer, just divide 1 by the integer, like in the examples.) For other negative powers, first find the reciprocal. After you find the reciprocal, raise it to the absolute value of the power. For example, 9^{-4} means the same thing as $\frac{1}{9^4}$.

Example 1. $11^{-1} = \frac{1}{11}$

Example 2. $6^{-2} = \frac{1}{6^2} = \frac{1}{6} \times \frac{1}{6} = \frac{1}{36}$

Example 3. $7^{-3} = \frac{1}{7^3} = \frac{1}{7 \times 7 \times 7} = \frac{1}{49 \times 7} = \frac{1}{343}$

Example 4. $8^{-1} = \frac{1}{8}$

1) $6^{-1} =$

2) $7^{-2} =$

3) $4^{-3} =$

4) $2^{-5} =$

5) $8^{-2} =$

6) $3^{-4} =$

7) $10^{-3} =$

8) $12^{-1} =$

9) $5^{-4} =$

10) $(-9)^{-2} =$

11) $2^{-7} =$

12) $(-3)^{-5} =$

13) $1^{-1} =$

14) $(-20)^{-1} =$

15) $8^{-3} =$

1.6 Square Roots

To find a square root, ask yourself, "Which number multiplied by itself is equal to the value under the radical?"

A square root has two possible answers. Why? A negative value squared is positive. Note that $13^2 = 13 \times 13 = 169$ and $(-13)^2 = -13 \times (-13) = 169$. Since 13^2 and $(-13)^2$ both equal 169, the square root of 169 is $\sqrt{169} = \pm 13$. The ± 13 is a shorthand way of saying that -13 and 13 are both answers to $\sqrt{169}$.

Example 1. $\sqrt{9} = \pm 3$ since $3^2 = 3 \times 3 = 9$ and since $(-3)^2 = -3 \times (-3) = 9$

Example 2. $\sqrt{121} = \pm 11$ since $11^2 = 11 \times 11 = 121$ and since $(-11)^2 = -11 \times (-11) = 121$

1) $\sqrt{16} =$

2) $\sqrt{4} =$

3) $\sqrt{225} =$

4) $\sqrt{25} =$

5) $\sqrt{100} =$

6) $\sqrt{64}$

7) $\sqrt{625} =$

8) $\sqrt{81} =$

9) $\sqrt{289} =$

10) $\sqrt{36} =$

11) $\sqrt{144} =$

12) $\sqrt{1} =$

13) $\sqrt{196} =$

14) $\sqrt{49} =$

1.7 Cube Roots

To find a cube root, ask yourself, "Which number cubed equals the value under the radical?"

The cube root of a positive number is positive, whereas the cube root of a negative number is negative, as shown in the examples.

Example 1. $\sqrt[3]{2197} = 13$ because $13^3 = 13 \times 13 \times 13 = 169 \times 13 = 2197$

Example 2. $\sqrt[3]{-8} = -2$ because $(-2)^3 = -2 \times (-2) \times (-2) = 4 \times (-2) = -8$

1) $\sqrt[3]{27} =$

2) $\sqrt[3]{125} =$

3) $\sqrt[3]{64} =$

4) $\sqrt[3]{-216} =$

5) $\sqrt[3]{729} =$

6) $\sqrt[3]{-343} =$

7) $\sqrt[3]{1000} =$

8) $\sqrt[3]{512} =$

1.8 Roots

To find a general root, ask yourself, "Which number raised to the indicated power equals the value under the radical?" It may help to review Sec. 1.3. In this section, we're basically doing Sec. 1.3 in reverse. Study the examples.

For an even root like $\sqrt[4]{81}$ (the 4 is even), use the $\pm$ sign to indicate two possible answers. For an odd root like $\sqrt[5]{32}$ (the 5 is odd), the answer has the same sign as the number under the radical.

Example 1. $\sqrt[4]{81} = \pm3$ because $(\pm3)^4 = 3 \times 3 \times 3 \times 3 = 9 \times 9 = 81$

Example 2. $\sqrt[5]{32} = 2$ because $2^5 = 2 \times 2 \times 2 \times 2 \times 2 = 4 \times 4 \times 2 = 16 \times 2 = 32$

Example 3. $\sqrt[5]{-32} = -2$ because $(-2)^5 = -32$

1) $\sqrt[4]{256} =$

2) $\sqrt[3]{512} =$

3) $\sqrt[4]{625} =$

4) $\sqrt[4]{16} =$

5) $\sqrt[5]{-243} =$

6) $\sqrt[5]{100,000} =$

7) $\sqrt[3]{-216} =$

8) $\sqrt[4]{4096} =$

9) $\sqrt[4]{10,000} =$

10) $\sqrt[6]{64} =$

11) $\sqrt[6]{729} =$

12) $\sqrt[5]{-1024} =$

13) $\sqrt[3]{343} =$

14) $\sqrt[9]{-512} =$

15) $\sqrt[4]{1296} =$

16) $\sqrt[3]{1,000,000} =$

17) $\sqrt[5]{-3125} =$

18) $\sqrt[3]{-8000} =$

1.9 Prime Factorization

A prime number is evenly divisible only by itself and one. Examples of prime numbers are: 2, 3, 5, 7, 11, 13, 17, and 19. The prime numbers that multiply together to make a number are the prime factorization of the number. For example, $2^3 \times 5$ is the prime factorization of 40 because $2^3 \times 5 = 2 \times 2 \times 2 \times 5 = 4 \times 10 = 40$. To determine the prime factorization of a number, ask yourself if the number is divisible by 2, 3, 5, 7, etc. Here are a few handy tips:

- If the number is even, it is divisible by 2. For example, 76 is even.
- If the digits add up to a multiple of 3, it is divisible by 3. For example, the digits of 132 add up to 6, which is a multiple of 3. Therefore, 132 is divisible by 3.
- If the number ends with 0 or 5, it is divisible by 5. For example, 845 is divisible by 5.

A simple way to determine the prime factorization of a number is to divide the number by a prime number, divide the result by a prime number, and so on until the result is itself prime. These prime numbers make the prime factorization. This is shown in the examples.

Example 1. What is the prime factorization of 63?

$$63 \div \boxed{7} = 9$$
$$9 \div \boxed{3} = \boxed{3}$$

The prime numbers are shown in boxes. Put these prime numbers together to determine that the prime factorization of 63 is $3^2 \times 7$ since $3^2 \times 7 = 9 \times 7 = 63$. Final answer: $3^2 \times 7$.

Example 2. What is the prime factorization of 210?

$$210 \div \boxed{7} = 30$$
$$30 \div \boxed{5} = 6$$
$$6 \div \boxed{3} = \boxed{2}$$

The prime factorization of 210 is $2 \times 3 \times 5 \times 7$ since $2 \times 3 \times 5 \times 7 = 6 \times 35 = 210$.

Example 3. What is the prime factorization of 100?

$$100 \div \boxed{5} = 20$$
$$20 \div \boxed{5} = 4$$
$$4 \div \boxed{2} = \boxed{2}$$

The prime factorization of 100 is $2^2 \times 5^2$ since $2^2 \times 5^2 = 4 \times 25 = 100$.

1) What is the prime factorization of 12?

2) What is the prime factorization of 45?

3) What is the prime factorization of 99?

4) What is the prime factorization of 70?

5) What is the prime factorization of 175?

6) What is the prime factorization of 196?

7) What is the prime factorization of 250?

8) What is the prime factorization of 72?

9) What is the prime factorization of 220?

1.10 Factor Out Perfect Squares

Consider the irrational number $\sqrt{18}$. The number 18 includes a perfect square in the sense that $18 = 9 \times 2$ and $9 = 3^2$. We call 9 a perfect square because $\sqrt{9} = 3$. Since 18 includes the perfect square 9, we can factor this perfect square out of the square root as follows:

$$\sqrt{18} = \sqrt{9 \times 2} = \sqrt{9} \times \sqrt{2} = 3 \times \sqrt{2} = 3\sqrt{2}$$

Note that $3\sqrt{2}$ is a shorthand way of writing $3 \times \sqrt{2}$. We used a rule that $\sqrt{xy} = \sqrt{x} \times \sqrt{y}$. You can check that this works. For example, compare $\sqrt{36} = 6$ to $\sqrt{4 \times 9} = \sqrt{4} \times \sqrt{9} = 2 \times 3 = 6$.

Example 1. $\sqrt{75} = \sqrt{25 \times 3} = \sqrt{25} \times \sqrt{3} = 5 \times \sqrt{3} = 5\sqrt{3}$

Example 2. $\sqrt{32} = \sqrt{16 \times 2} = \sqrt{16} \times \sqrt{2} = 4 \times \sqrt{2} = 4\sqrt{2}$

Example 3. $\sqrt{245} = \sqrt{49 \times 5} = \sqrt{49} \times \sqrt{5} = 7 \times \sqrt{5} = 7\sqrt{5}$

1) $\sqrt{12} =$

2) $\sqrt{63} =$

3) $\sqrt{80} =$

4) $\sqrt{108} =$

5) $\sqrt{27} =$

6) $\sqrt{128} =$

7) $\sqrt{175} =$

8) $\sqrt{48} =$

9) $\sqrt{243} =$

10) $\sqrt{44} =$

11) $\sqrt{500} =$

12) $\sqrt{98} =$

13) $\sqrt{288} =$

14) $\sqrt{320} =$

1.11 Multiplying Square Roots

A square root multiplied by itself simply removes the radical sign. For example, $\sqrt{5} \times \sqrt{5} = 5$. You can easily check this rule for a perfect square. For example, compare $\sqrt{9} \times \sqrt{9} = 9$ with $\sqrt{9} \times \sqrt{9} = 3 \times 3 = 9$.

If a square root is raised to a power, apply the previous rule to each pair of square roots. For example, $\left(\sqrt{3}\right)^6 = \sqrt{3} \times \sqrt{3} \times \sqrt{3} \times \sqrt{3} \times \sqrt{3} \times \sqrt{3} = 3 \times 3 \times 3 = 27$.

If the power is an odd number, there will be one square root left over. For example, $\left(\sqrt{2}\right)^3 = \sqrt{2} \times \sqrt{2} \times \sqrt{2} = 2 \times \sqrt{2} = 2\sqrt{2}$.

Example 1. $\sqrt{11} \times \sqrt{11} = 11$

Example 2. $\left(\sqrt{7}\right)^4 = \sqrt{7} \times \sqrt{7} \times \sqrt{7} \times \sqrt{7} = 7 \times 7 = 49$

Example 3. $\left(\sqrt{19}\right)^5 = \sqrt{19} \times \sqrt{19} \times \sqrt{19} \times \sqrt{19} \times \sqrt{19} = 19 \times 19 \times \sqrt{19} = 361 \times \sqrt{19} = 361\sqrt{19}$

1) $\sqrt{6} \times \sqrt{6} =$

2) $\left(\sqrt{5}\right)^3 =$

3) $\left(\sqrt{10}\right)^4 =$

4) $\left(\sqrt{3}\right)^3 =$

5) $\left(\sqrt{2}\right)^5 =$

6) $\left(\sqrt{13}\right)^2 =$

7) $\left(\sqrt{11}\right)^4 =$

8) $\left(\sqrt{5}\right)^5 =$

9) $\left(\sqrt{2}\right)^9 =$

10) $\left(\sqrt{3}\right)^8 =$

11) $\sqrt{7} \times \sqrt{7} =$

12) $\left(\sqrt{6}\right)^6 =$

13) $\left(\sqrt{11}\right)^3 =$

14) $\left(\sqrt{15}\right)^4 =$

15) $\left(\sqrt{7}\right)^5 =$

16) $\left(\sqrt{10}\right)^6 =$

2 ORDER OF OPERATIONS

Consider the expression $8 - 3 \times 2$. Does the **order of operations** matter? Put another way, do you get the same answer when you subtract first and then multiply as when you multiply first and then subtract? Test it out!

- If we subtract $8 - 3 = 5$ first, we get $5 \times 2 = 10$.
- If we multiply $3 \times 2 = 6$ first, we get $8 - 6 = 2$.

If we subtract before we multiply, the answer is 10, but if we multiply before we subtract, the answer is 2. Therefore, the order of operations **does** matter. It can affect the answer. We will learn in Sec.'s 2.6-2.8 that the convention is to multiply before subtracting (such that the correct solution to $8 - 3 \times 2$ is $8 - 6 = 2$).

Can you think of any situations where order doesn't matter? Think about it.

Consider the expression $2 + 5 + 3$. Does the order of operations matter for this? Test it out.

- If we add $2 + 5$ first, we get $7 + 3 = 10$.
- If we add $5 + 3$ first, we get $2 + 8 = 10$.

Whether we add $2 + 5$ first or add $5 + 3$ first, either way the answer is 10. By trying other possibilities, you can easily convince yourself that if you're only adding numbers, it doesn't matter in which order you add them. The property $(a + b) + c = a + (b + c)$ is referred to as the **associative property** of addition. (There is a similar property of multiplication.)

There is another way where order doesn't matter for addition. Do $4 + 8$ and $8 + 4$ equal the same value? Yes. Both equal 12. The property $a + b = b + a$ is referred to as the **commutative property** of addition. (There is a similar property for multiplication.)

Is subtraction commutative? That is, do $a - b$ and $b - a$ equal the same value? No. Subtraction isn't commutative. For example, $9 - 6 = 3$ whereas $6 - 9 = -3$. That minus sign makes a big difference. If you don't think that minus sign is a big deal, ask yourself this question: Would you rather have somebody owe you \$3, or would you rather owe \$3 to someone else. When we put it like this, you should see that it does matter.

Similarly, division isn't commutative. For example, $8 \div 4 = 2$ whereas $4 \div 8 = \frac{1}{2}$.

Another interesting property of arithmetic is the **distributive property**, which involves both multiplication and addition: $a \times (b + c) = a \times b + a \times c$. On the left-hand side, we must add the values in parentheses before we multiply. Let's try it using $a = 4$, $b = 3$, and $c = 2$. The left-hand side is $4 \times (3 + 2) = 4 \times 5 = 20$ and the right-hand side is $4 \times 3 + 4 \times 2 = 12 + 8 = 20$. If you try other values, you should be convinced that the distributive property holds.

2.1 Addition

The associative property of addition is $(a + b) + c = a + (b + c)$. What this means is that it doesn't matter in which order you add three (or more) numbers together. For example, you can see that $(4 + 2) + 3 = 6 + 3 = 9$ is equivalent to $4 + (2 + 3) = 4 + 5 = 9$.

We can use the associative property to add multi-digit numbers. For example:
$$28 + 33 = 20 + 8 + 30 + 3 = (20 + 30) + (8 + 3) = 50 + 11 = 61$$

Example 1. $47 + 8 = 40 + 7 + 8 = 40 + (7 + 8) = 40 + 15 = 55$

Example 2. $85 + 75 = 80 + 5 + 70 + 5 = (80 + 70) + (5 + 5) = 150 + 10 = 160$

1) $29 + 7 =$ 36

2) $13 + 13 =$ 26

3) $22 + 19 =$ 41

4) $44 + 31 =$ 75

5) $39 + 22 =$ 61

6) $16 + 73 =$ 89

7) $45 + 45 =$ 90

8) $68 + 76 =$ 144

9) $74 + 68 =$ 142

10) $87 + 79 =$ 166

11) $16 + 92 =$ 108

12) $99 + 99 =$ 198

When you add a positive number to a negative number, the result is basically subtraction, but you must take care with the sign. If the absolute value of the negative number is greater than the positive number, the answer will be negative. If the absolute value of the negative number is less than the positive number, the answer will be positive.

When you add two negative numbers together, first add the positive values together, and then put a minus sign in front. Adding two negative numbers results in a more negative number.

Example 3. $8 + (-5) = 8 - 5 = 3$

Example 5. $-7 + 16 = 9$

Example 7. $-4 + (-8) = -(4 + 8) = -12$

Example 4. $5 + (-8) = 5 - 8 = -3$

Example 6. $-16 + 7 = -9$

Example 8. $-2 + (-2) = -(2 + 2) = -4$

13) $14 + (-7) = 7$

14) $4 + (-9) = -5$

15) $-2 + (-6) = -8$

16) $-8 + 15 = 7$

17) $7 + (-8) = -1$

18) $6 + (-7) = -1$

19) $3 + (-5) = -2$

20) $-5 + (-9) = -14$

21) $-6 + 21 = 15$

22) $-8 + 8 = 0$

23) $-14 + 9 = -24$

24) $-9 + (-9) = -18$

25) $-7 + (-8) = -15$

26) $-37 + 22 = -15$

27) $36 + (-6) = 30$

28) $-11 + 5 = -6$

29) $-6 + (-8) = -14$

30) $17 + (-19) = -2$

31) $-56 + 8 = -48$

32) $-7 + 7 = 0$

33) $-3 + (-3) = -6$

34) $18 + (-9) = 97$

35) $-9 + (-7) = -16$

36) $55 + (-88) = 143$

2.2 Subtraction

Subtraction can result in a negative answer. For example, $4 - 7 = -3$ whereas $7 - 4 = 3$. If you subtract more than a number has, the answer is negative.

Subtracting a negative number from a positive number is equivalent to addition. For example, $4 - (-7) = 4 + 7 = 11$.

Subtracting a positive number from a negative number makes a more negative number. Add the values together and put a minus sign in front: $-4 - 7 = -(4 + 7) = -11$.

Subtracting one negative number from another equates to adding a positive number to a negative number, which is similar to Sec. 2.1. For example, $-4 - (-7) = -4 + 7 = 3$.

Example 1. $9 - 3 = 6$

Example 3. $9 - (-3) = 9 + 3 = 12$

Example 5. $-9 - (-3) = -9 + 3 = -6$

Example 2. $3 - 9 = -6$

Example 4. $-9 - 3 = -(9 + 3) = -12$

Example 6. $-3 - (-9) = -3 + 9 = 6$

1) $5 - 8 =$

2) $-8 - 2 =$

3) $4 - 9 =$

4) $11 - (-4) =$

5) $8 - 7 =$

6) $-7 - 3 =$

7) $-5 - (-8) =$

8) $-10 - (-7) =$

9) $6 - 20 =$

10) $32 - 9 =$

11) $-7 - (-12) =$

12) $-8 - 13 =$

13) $14 - (-5) =$

14) $0 - (-6) =$

15) $-6 - 6 =$

16) $19 - (-8) =$

17) $99 - 33 =$

18) $-9 - (-9) =$

19) $0 - 9 =$

20) $-6 - (-43) =$

2.3 Multiplication

When multiplying numbers, an even number of minus signs results in a positive answer, but an odd number of minus signs results in a negative answer.

Example 1. $4 \times 5 = 20$ **Example 2.** $4 \times (-5) = -20$

Example 3. $-4 \times 5 = -20$ **Example 4.** $-4 \times (-5) = 20$

1) $3 \times (-9) =$ 2) $-6 \times (-8) =$

3) $-4 \times 4 =$ 4) $5 \times 6 =$

5) $-5 \times (-8) =$ 6) $6 \times (-7) =$

7) $7 \times 8 =$ 8) $-3 \times 9 =$

9) $6 \times (-9) =$ 10) $-8 \times (-8) =$

11) $-4 \times 8 =$ 12) $7 \times 9 =$

13) $6 \times 7 =$ 14) $-3 \times 8 =$

15) $-7 \times (-9) =$ 16) $5 \times (-7) =$

17) $-4 \times 7 =$ 18) $-9 \times (-9) =$

19) $3 \times 9 =$ 20) $6 \times (-5) =$

21) $-4 \times (-8) =$ 22) $7 \times 8 =$

23) $-4 \times 6 =$ 24) $3 \times (-5) =$

25) $5 \times (-9) =$ 26) $-7 \times (-9) =$

27) $6 \times 9 =$ 28) $-8 \times 9 =$

The distributive property is $a \times (b + c) = a \times b + a \times c$. We can use the distributive property to multiply a single-digit number and two-digit number together. For example:
$$4 \times 17 = 4 \times (10 + 7) = 4 \times 10 + 4 \times 7 = 40 + 28 = 68$$
To multiply a two-digit number and a single-digit number together, note that we could also write the distributive property as $(b + c) \times a = b \times a + c \times a$.

Example 5. $6 \times 12 = 6 \times (10 + 2) = 6 \times 10 + 6 \times 2 = 60 + 12 = 72$

Example 6. $34 \times 5 = (30 + 4) \times 5 = 30 \times 5 + 4 \times 5 = 150 + 20 = 170$

29) $24 \times 3 =$

30) $7 \times 11 =$

31) $26 \times 7 =$

32) $5 \times 12 =$

33) $-6 \times 14 =$

34) $13 \times 6 =$

35) $14 \times 7 =$

36) $7 \times 19 =$

37) $4 \times (-32) =$

38) $22 \times 6 =$

39) $9 \times 41 =$

40) $26 \times 8 =$

41) $-5 \times (-15) =$

42) $32 \times 6 =$

1\5  **2.4 Division**

Some division problems have remainders. For example, $23 \div 4 = 5R3$ (read this as "5 with a remainder of 3"). Why? 23 isn't evenly divisible by 4. The largest number no greater than 23 that is evenly divisible by 4 is 20. Since $23 - 20 = 3$, the remainder is 3. (The 5 comes from $20 \div 4 = 5$.) Try to follow and understand the logic behind the examples.

detraf ohw   W

Example 1. $15 \div 5 = 3$ (there is no remainder since 15 is evenly divisible by 5)

Example 2. $20 \div 6 = (18 \div 6)R(20 - 18) = 3R2$ (3 with a remainder of 2)

Example 3. $7 \div 3 = (6 \div 3)R(7 - 6) = 2R1$ (2 with a remainder of 1)

Example 4. $39 \div 7 = (35 \div 7)R(39 - 35) = 5R4$ (5 with a remainder of 4)

1) $28 \div 4 = 7$

2) $58 \div 9 = 6\ r\ 4$

$4\overline{)23}$ with 5 above, 20 below

3) $21 \div 3 = 7$

4) $36 \div 6 = 6$

$9\overline{)58}$ with 6 above, 54 below, 4 remainder

5) $19 \div 5 = 3\ r\ 4$ $5\overline{)19}$ with 3 above, 15 below, r 4

6) $16 \div 4 = 4$

7) $56 \div 8 = 7$ 4

8) $81 \div 9 = 9$

9) $63 \div 7 = 9$

10) $30 \div 5 = 6$

11) $41 \div 6 = 6\ r\ 5$ $6\overline{)41}$ with 36 below, 5 remainder

12) $49 \div 7 = 7$

13) $18 \div 6 = 3$

14) $54 \div 9 = 6$

15) $56 \div 7 = 8$

16) $29 \div 2 = 14\ r\ 1$ $2\overline{)29}$ with 14 above, 28 below, 1 remainder

17) $32 \div 8 = 4$

18) $48 \div 6 = 8$

19) $66 \div 8 = 8\ r\ 2$ $8\overline{)66}$ with 8 above, 64 below, 2 remainder, r 2

20) $63 \div 9 = 7$

21) $36 \div 4 = 9$

22) $45 \div 7 = 6\ r\ 3$ $7\overline{)45}$ with 6 above, 42 below, 3 remainder

23) $72 \div 9 = 8$

24) $35 \div 5 = 7$

WTF
what the fudge

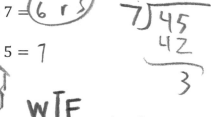

When dividing numbers, an even number of minus signs results in a positive answer, but an odd number of minus signs results in a negative answer.

Example 5. $18 \div 3 = 6$ **Example 6.** $18 \div (-3) = -6$

Example 7. $-18 \div 3 = -6$ **Example 8.** $-18 \div (-3) = 6$

25) $24 \div (-6) =$ -4

26) $-16 \div 8 =$ -2

27) $25 \div 5 =$ 5

28) $-27 \div (-3) =$ 9

29) $-36 \div 6 =$ -6

30) $12 \div (-4) =$ -4

31) $-45 \div (-9) =$ 5

32) $42 \div 7 =$ 6

33) $54 \div (-6) =$ -9

34) $-32 \div 8 =$ -4

35) $20 \div 5 =$ 4

36) $-40 \div (-8) =$ 5

37) $-44 \div 4 =$ -11

38) $60 \div (-10) =$ -6

39) $-25 \div (-5) =$ 5

40) $99 \div 9 =$ 11

41) $60 \div 5 =$ 6

42) $64 \div (-8) =$ -8

43) $-63 \div 7 =$ -9

44) $-36 \div (-9) =$ 4

45) $50 \div (-10) =$ -5

46) $-20 \div (-4) =$ 5

47) $100 \div 10 =$ 10

48) $-72 \div 8 =$ -9

49) $-35 \div (-5) =$ 7

50) $48 \div (-8) =$ -6

51) $-56 \div 7 =$ -8

52) $90 \div 9 =$ 10

2.5 Properties of Arithmetic Operators

The commutative property of addition and the commutative property of multiplication state that the order in which two numbers are added or multiplied doesn't matter.

$$a + b = b + a \quad \text{(commutative property of addition)}$$

$$a \times b = b \times a \quad \text{(commutative property of multiplication)}$$

The associative property of addition and the associative property of multiplication state that the order in which three numbers are added or multiplied doesn't matter. That is, it doesn't matter which pair of numbers you add or multiply first.

$$(a + b) + c = a + (b + c) \quad \text{(associative property of addition)}$$

$$(a \times b) \times c = a \times (b \times c) \quad \text{(associative property of multiplication)}$$

The distributive property combines the operations of addition and multiplication together:

$$a \times (b + c) = a \times b + a \times c \quad \text{(distributive property)}$$

The identity property of addition states that adding zero has no effect. The identity property of multiplication states that multiplying by one has no effect.

$$a + 0 = 0 + a = a \quad \text{(identity property of addition)}$$

$$a \times 1 = 1 \times a = a \quad \text{(identity property of multiplication)}$$

The inverse property of addition states that subtraction (or the addition of a negative value) is the opposite of addition in the sense that a number plus its negative counterpart is equal to zero. The inverse property of multiplication states that division (or the multiplication by a reciprocal) is the opposite of multiplication in the sense that a number times its reciprocal is equal to one. (This is why a reciprocal is called a multiplicative inverse.)

$$a + (-a) = a - a = 0 \quad \text{(inverse property of addition)}$$

$$a \times \frac{1}{a} = a \times a^{-1} = a \div a = \frac{a}{a} = 1 \quad \text{(inverse property of multiplication)}$$

Example 1. $3 \times 2 = 2 \times 3$ applies the commutative property of multiplication

Example 2. $0 - 1 = -1$ applies the identity property of addition (with $a = -1$)

1) Which property is applied in $5 - 8 = -8 + 5$?

2) Which property is applied in $(4 - 2) \times 3 = 4 \times 3 - 2 \times 3$?

3) Which property is applied in $\frac{1}{6} \times 6 = 1$?

2.6 Combining Operations

When an expression contains multiple operations, it is customary to do multiplication and division before doing addition and subtraction. First do multiplication and division working from left to right. Once you finish multiplying and dividing, then do addition and subtraction working from left to right. (If a problem wanted you to do addition or subtraction sooner, the problem would include parentheses, as we will learn in Sec. 2.8.)

For example, consider the expression below.

$$9 - 6 \div 3 + 2 \times 5$$

According to the rules, it would be incorrect to subtract 6 from 9 first. According to the rules, we must carry out multiplication and division first starting from the left. It is correct to begin with $6 \div 3 = 2$ and $2 \times 5 = 10$. If we do these first, we get:

$$9 - 6 \div 3 + 2 \times 5 = 9 - 2 + 10$$

Now that we have finished multiplying and dividing, we may add and subtract from left to right:

$$9 - 2 + 10 = 7 + 10 = 17$$

Beware that if you don't follow the rules, you will get the wrong answer.

Example 1. $3 + 2 \times 5 - 4 \div 2 = 3 + 10 - 2 = 13 - 2 = 11$

Remember: Multiply and divide (left to right) **before** you add and subtract (left to right).

Example 2. $5 + 12 \div 2 \times 3 = 5 + 6 \times 3 = 5 + 18 = 23$

Note: The rules **don't** say to multiply before you divide. They say to do both multiplication and division from left to right before you add and subtract.

Example 3. $3 \times 6 - 6 \div 2 = 18 - 3 = 15$

1) $16 - 8 \div 2 + 2 =$

2) $24 \div 3 \times 2 =$

3) $5 + 3 - 2 + 9 - 7 =$

4) $6 + 3 \times 7 + 18 \div 3 =$

5) $4 \times 8 - 2 \times 9 + 12 \div 4 =$

6) $36 \div 6 \times 2 - 8 \times 9 \div 4 =$

7) $5 + 5 \times 5 \div 5 - 5 =$

8) $56 \div 7 + 3 \times 8 - 6 \times 7 =$

9) $3 + 6 \times 9 - 4 \times 5 + 48 \div 8 =$

10) $9 + 8 - 40 \div 5 \times 2 - 24 \div 6 \div 4 =$

11) $4 \times 7 - 5 \times 5 + 18 \div 2 - 36 \div 9 =$

12) $2 \div 2 + 2 \times 2 \times 2 - 2 \times 2 + 2 =$

13) $81 \div 9 \div 3 + 6 \times 4 \div 3 - 42 \div 7 + 3 \times 3 =$

14) $9 \times 8 - 7 \times 6 - 5 \times 4 - 3 \times 2 - 1 \times 0 =$

15) $7 + 3 \times 5 + 6 \times 6 \div 4 - 2 \times 3 \times 4 + 8 =$

2.7 Operations with Exponents

If an expression includes exponents or roots, you must deal with the exponents and roots before you multiply and divide.

- First deal with any exponents or roots.
- Next multiply and divide from left to right.
- Finally, add and subtract from left to right.

Example 1. $2 \times 5^2 = 2 \times 25 = 50$

Example 2. $4 + 2^3 \times 3^2 = 4 + 8 \times 9 = 4 + 72 = 76$

Remember: Exponents and roots first, then multiply and divide, and then add and subtract.

Example 3. $\sqrt[3]{8} + \sqrt[3]{27} \times \sqrt[3]{64} = 2 + 3 \times 4 = 2 + 12 = 14$

1) $15 - 3^2 + 2^4 =$ 22

2) $8 - 4^2 \div 2^2 =$ ~~8~~ 4

3) $6 + 2 \times 5^2 - 7^2 =$ ~~8~~ ~~12~~ 7

4) $9^2 - 3 \times 5^2 + 24 \div 6 =$ ~~N~~ 10

5) $4^3 + 6^2 - 2^2 \times 5^2 =$ ~~60~~ 0

6) $\sqrt[5]{32} \times 3 + 8 =$???

7) $8^2 \div 4^2 \times 3^2 \div 2^2 =$ 9

2.8 Operations with Parentheses and Exponents

The acronym PEMDAS can help you remember the order of operations:

- P stands for parentheses. Simplify parentheses first.
- E stands for exponents. Deal with exponents and roots after parentheses.
- MD stands for multiplication and division. Working left to right, multiply and divide after dealing with exponents and roots (and before adding and subtracting).
- AS stands for addition and subtraction. Working left to right, add and subtract (after multiplying and dividing).

Example 1. $7 + (8 - 3) \times 2^2 = 7 + 5 \times 2^2 = 7 + 5 \times 4 = 7 + 20 = 27$

Remember: Parentheses first, then exponents, then multiply and divide, and then add and subtract.

Example 2. $(2 + 8 \div 4)^2 - 3 \times 3 = (2 + 2)^2 - 3 \times 3 = 4^2 - 3 \times 3 = 16 - 3 \times 3 = 16 - 9 = 7$

Note: Within parentheses, we still multiply and divide before we add and subtract.

1) $9 - (3 + 5) \div 2 =$ 5

2) $(12 - 5) \times (7 - 4) =$ 21

3) $3 \times (9 - 2^2) - 3^2 =$ 6

4) $(4 + 2 \times 3) \div 2 + 3 =$ 8

5) $(4 \times 6 - 2 \times 9)^2 = 36$

6) $(-3)^2 \div (-3) = 3$

7) $(5^2 - 4^2)^2 = 81$

8) $(3 - 8) \times 4 + 4^2 = 4$

9) $-3 - 4 \times (1 + 2) = 1576$

10) $-2 - 2^2 - (-2)^2 = 10$

11) $(7 - 5 \times 2)^3 - (4 - 2 \times 3)^2 = 31$

12) $(2 + 3) \times (9 - 5) \div (8 - 6) = 10$

13) $2^{9 - 3 \times (6 - 4)} =$

14) $\sqrt[3]{64} - 8 \times (11 - 5) + 11 =$

2.9 Why the Order of Operations Matters

Consider the expression $5 + 3 \times 2$. According to the rules, you should multiply $3 \times 2 = 6$ first to get $5 + 6 = 11$. Note that if instead you add $5 + 3 = 8$ first, you would get $8 \times 2 = 16$. Since 11 and 16 are different, you can see that the order does matter.

What if you wanted to make a problem where you add first? You could use parentheses to force addition first. Compare the two problems below.

$$(5 + 3) \times 2 = 8 \times 2 = 16$$
$$5 + 3 \times 2 = 5 + 6 = 11$$

Example 1. Calculate $(8 - 3) \times 2$ with and without parentheses.
- With parentheses: $(8 - 3) \times 2 = 5 \times 2 = 10$
- Without parentheses: $8 - 3 \times 2 = 8 - 6 = 2$

1) Calculate $24 \div (4 + 2)$ with and without parentheses.

2) Calculate $(6 + 8) \div 2$ with and without parentheses.

3) Calculate $3 \times (3 + 3)$ with and without parentheses.

4) Calculate $(15 - 6) \times 3$ with and without parentheses.

5) Calculate $(4 \times 9) \div (3 \times 4)$ with and without parentheses.

2.10 Parentheses Challenge

These exercises challenge you to insert one or more pairs of parentheses in such a way as to make a given expression equal a specified answer. Try out different possibilities.

Example 1. Add parentheses to make $5 \times 3 - 8 \times 2$ equal 14.
Answer: $(5 \times 3 - 8) \times 2 = (15 - 8) \times 2 = 7 \times 2 = 14$
Example 2. Add parentheses to make $2 + 3 \times 5 - 1$ equal 20.
Answer: $(2 + 3) \times (5 - 1) = 5 \times 4 = 20$

1) Add parentheses to make $3 \times 4 + 2 \times 5$ equal 90.

2) Add parentheses to make $72 \div 4 \times 2 - 6$ equal 3.

3) Add parentheses to make $4 + 4 \times 4 - 4$ equal 4.

4) Add parentheses to make $5 + 3 \times 9 - 2$ equal 56.

5) Add parentheses to make $5 \times 6 - 3 \times 8 + 8 \div 2$ equal 7.

6) Add parentheses to make $13 - 4 \times 2 + 3 - 7 \times 6$ equal 3.

3 FRACTIONS

The top number of a fraction is called a **numerator**, and the bottom number of a fraction is called a **denominator**. A fraction is **proper** if the numerator is smaller than the denominator (like $\frac{2}{3}$), and **improper** if the numerator is greater than the denominator (like $\frac{3}{2}$). Don't let the terms proper and improper mislead you; improper fractions are common in advanced work. A **mixed number** includes an integer along with a proper fraction (like $2\frac{1}{3}$, which means the same as 2 plus $\frac{1}{3}$).

Two different fractions can actually be equivalent. For example, $\frac{2}{4}$, $\frac{3}{6}$, and $\frac{5}{10}$ are all equivalent to $\frac{1}{2}$ because in each case the numerator is one-half as large as the denominator. The **reduced** form of a fraction is the form where the numerator and denominator don't share a common factor. When the numerator and denominator share a common factor, it is possible to reduce the fraction by dividing the numerator and denominator each by the **greatest common factor** (GCF). For example, $\frac{8}{12}$ can be reduced by dividing 8 and 12 each by 4 (since the GCF of 8 and 12 is 4): $\frac{8}{12} = \frac{8 \div 4}{12 \div 4} = \frac{2}{3}$.

The way to add or subtract two fractions is to make a **common denominator**. For two fractions, the **lowest common denominator** (LCD) is the **least common multiple** (LCM) that can be made from their denominators. For example, to add $\frac{5}{6}$ to $\frac{4}{9}$, the LCD is 18 (since the multiples of 6 are 6, 12, 18, 24, etc. and the multiples of 9 are 9, 18, 27, 36, etc.; note that 18 is the smallest multiple that 6 and 9 have in common). The fractions $\frac{5}{6}$ and $\frac{4}{9}$ can be changed into their LCD as follows: Multiply $\frac{5}{6}$ by $\frac{3}{3}$ to get $\frac{5 \times 3}{6 \times 3} = \frac{15}{18}$ and multiply $\frac{4}{9}$ by $\frac{2}{2}$ to get $\frac{4 \times 2}{9 \times 2} = \frac{8}{18}$. Now the fractions can be added: $\frac{5}{6} + \frac{4}{9} = \frac{15}{18} + \frac{8}{18} = \frac{15+8}{18} = \frac{23}{18}$.

It turns out to be easier to multiply fractions than to add or subtract them. Why? You don't need to find a common denominator to multiply fractions. For example, $\frac{3}{4} \times \frac{3}{5} = \frac{3 \times 3}{4 \times 5} = \frac{9}{20}$.

To divide two fractions, first you need to know how to take a reciprocal. To find the **reciprocal** of a fraction, simply swap the numerator and denominator. For example, the reciprocal of $\frac{2}{3}$ is $\frac{3}{2}$. Dividing two fractions is equivalent to multiplying the first fraction by the reciprocal of the second fraction. For example, $\frac{5}{7} \div \frac{2}{3} = \frac{5}{7} \times \frac{3}{2} = \frac{5 \times 3}{7 \times 2} = \frac{15}{14}$.

3.1 Reducing Fractions

To reduce a fraction, divide the numerator and denominator each by the greatest common factor (GCF). For example, the 9 and 12 of $\frac{9}{12}$ are both multiples of 3, allowing us to write:

$$\frac{9}{12} = \frac{9 \div 3}{12 \div 3} = \frac{3}{4}$$

If you divide by a common factor that isn't the GCF, you'll need to divide by another common factor in order to completely reduce the fraction. For example, consider $\frac{24 \div 6}{36 \div 6} = \frac{4}{6}$ which can be reduced further by $\frac{4 \div 2}{6 \div 2} = \frac{2}{3}$. We could achieve this in one step with $\frac{24 \div 12}{36 \div 12} = \frac{2}{3}$.

Example 1. $\frac{15}{25} = \frac{15 \div 5}{25 \div 5} = \frac{3}{5}$

Example 2. $\frac{24}{6} = \frac{24 \div 6}{6 \div 6} = \frac{4}{1} = 4$

Note that the fraction line equates to division. That is, $\frac{24}{6} = 24 \div 6 = 4$ and $\frac{4}{1} = 4 \div 1 = 4$.

1) $\frac{8}{12} =$

2) $\frac{20}{40} =$

3) $\frac{9}{21} =$

4) $\frac{25}{10} =$

5) $\frac{10}{12} =$

6) $\frac{8}{32} =$

7) $\frac{18}{24} =$

8) $\frac{16}{28} =$

9) $\frac{32}{18} =$

10) $\frac{9}{15} =$

11) $\frac{14}{49} =$

12) $\frac{45}{72} =$

13) $\frac{10}{25} =$

14) $\frac{36}{8} =$

15) $\frac{42}{7} =$

16) $\frac{24}{16} =$

17) $\frac{45}{18} =$

18) $\frac{32}{48} =$

3.2 Mixed Numbers

A mixed number like $4\frac{3}{5}$ can be converted into an improper fraction as follows:

- Multiply the integer (4) by the denominator (5). In this case, $4 \times 5 = 20$.
- Add this to the old numerator (3) to get the new numerator. In this case, $20 + 3 = 23$.
- Put the new numerator over the old denominator: In this case, it is $\frac{23}{5}$.

$$4\frac{3}{5} = \frac{4 \times 5 + 3}{5} = \frac{23}{5}$$

Example 1. $2\frac{1}{3} = \frac{2 \times 3 + 1}{3} = \frac{7}{3}$

Example 2. $5\frac{2}{7} = \frac{5 \times 7 + 2}{7} = \frac{37}{7}$

1) $3\frac{4}{9} =$

2) $6\frac{3}{4} =$

3) $2\frac{6}{7} =$

4) $5\frac{3}{5} =$

5) $7\frac{2}{5} =$

6) $9\frac{1}{6} =$

7) $4\frac{7}{8} =$

8) $5\frac{5}{9} =$

9) $8\frac{3}{5} =$

10) $2\frac{7}{8} =$

11) $6\frac{4}{7} =$

12) $7\frac{2}{9} =$

13) $4\frac{4}{5} =$

14) $8\frac{5}{6} =$

15) $5\frac{6}{7} =$

16) $3\frac{3}{8} =$

17) $9\frac{5}{6} =$

18) $7\frac{7}{9} =$

19) $6\frac{7}{8} =$

20) $9\frac{8}{9} =$

To convert an improper fraction into a mixed number:

- Think of it as the numerator divided by the denominator. For example, $\frac{27}{4} = 27 \div 4$.

- Figure out the answer to the division problem, including the remainder (Sec. 2.4). For example, $27 \div 4 = 6R3$ (that is, 6 with a remainder of 3).

- Write this as an integer plus the remainder divided by the original denominator. In this example, $6R3 = 6\frac{3}{4}$ (since $\frac{27}{4}$ had a denominator of 4).

$$\frac{27}{4} = 27 \div 4 = 6R3 = 6\frac{3}{4}$$

Example 3. $\frac{11}{3} = 11 \div 3 = 3R2 = 3\frac{2}{3}$ **Example 4.** $\frac{23}{5} = 23 \div 5 = 4R3 = 4\frac{3}{5}$

21) $\frac{7}{2} =$

22) $\frac{20}{3} =$

23) $\frac{23}{6} =$

24) $\frac{19}{7} =$

25) $\frac{35}{8} =$

26) $\frac{29}{5} =$

27) $\frac{34}{9} =$

28) $\frac{44}{7} =$

29) $\frac{39}{4} =$

30) $\frac{45}{8} =$

31) $\frac{29}{6} =$

32) $\frac{26}{3} =$

33) $\frac{49}{6} =$

34) $\frac{31}{4} =$

35) $\frac{68}{7} =$

36) $\frac{16}{9} =$

37) $\frac{71}{8} =$

38) $\frac{52}{7} =$

39) $\frac{41}{6} =$

40) $\frac{89}{9} =$

3.3 Adding and Subtracting Fractions

To add or subtract fractions, multiply the numerator and denominator of each fraction by the number needed to express each fraction with a common denominator (ideally, the lowest common denominator). For example, the denominators are 6 and 9 in $\frac{5}{6} - \frac{4}{9}$. The least common multiple of 6 and 9 is 18 (found by comparing 6, 12, 18, 24, etc. with 9, 18, 27, etc.). Multiply $\frac{5}{6}$ by $\frac{3}{3}$ and multiply $\frac{4}{9}$ by $\frac{2}{2}$ in order to make a common denominator of 18. (This is the lowest common denominator because 18 is the least common multiple of 6 and 9.) Once the fractions have the same denominator, we can simply subtract the numerators.

$$\frac{5}{6} - \frac{4}{9} = \frac{5 \times 3}{6 \times 3} - \frac{4 \times 2}{9 \times 2} = \frac{15}{18} - \frac{8}{18} = \frac{15 - 8}{18} = \frac{7}{18}$$

Example 1. $\frac{8}{15} + \frac{7}{20} = \frac{8\times4}{15\times4} + \frac{7\times3}{20\times3} = \frac{32}{60} + \frac{21}{60} = \frac{32+21}{60} = \frac{53}{60}$

Example 2. $\frac{11}{12} - \frac{3}{4} = \frac{11}{12} - \frac{3\times3}{4\times3} = \frac{11}{12} - \frac{9}{12} = \frac{11-9}{12} = \frac{2}{12} = \frac{2\div2}{12\div2} = \frac{1}{6}$

Note: Since $\frac{2}{12}$ is reducible, we reduced the answer to $\frac{1}{6}$ using the technique from Sec. 3.1.

Example 3. $\frac{3}{5} + \frac{1}{6} = \frac{3\times6}{5\times6} + \frac{1\times5}{6\times5} = \frac{18}{30} + \frac{5}{30} = \frac{18+5}{30} = \frac{23}{30}$

Example 4. $\frac{16}{9} - \frac{5}{12} = \frac{16\times4}{9\times4} - \frac{5\times3}{12\times3} = \frac{64}{36} - \frac{15}{36} = \frac{64-15}{36} = \frac{49}{36}$

1) $\frac{3}{4} + \frac{1}{6} =$

2) $\frac{5}{6} - \frac{3}{8} =$

3) $\frac{2}{3} + \frac{3}{5} =$

4) $\frac{7}{6} - \frac{2}{3} =$

5) $\frac{7}{10} + \frac{4}{15} =$

6) $\frac{11}{12} - \frac{1}{4} =$

7) $\frac{3}{2} + \frac{2}{3} =$

8) $\frac{9}{8} - \frac{5}{12} =$

9) $\frac{7}{12} + \frac{5}{18} =$

10) $\frac{8}{15} - \frac{2}{9} =$

3.4 Multiplying Fractions

You don't need a common denominator to multiply fractions. Just multiply their numerators together and multiply their denominators together.

$$\frac{2}{3} \times \frac{9}{8} = \frac{2 \times 9}{3 \times 8} = \frac{18}{24} = \frac{18 \div 6}{24 \div 6} = \frac{3}{4}$$

As with the above example, the answer can sometimes be reduced using the technique from Sec. 3.1. There is an alternative way to multiply the fractions above. Compare the solution below to the one above. We applied the commutative property of multiplication (that order doesn't matter when multiplying numbers), and observed that $\frac{2}{8}$ reduces to $\frac{1}{4}$ and that $\frac{9}{3} = 3$. It doesn't matter which of these methods you use. As you can see, the answer is the same.

$$\frac{2}{3} \times \frac{9}{8} = \frac{2 \times 9}{3 \times 8} = \frac{2 \times 9}{8 \times 3} = \frac{2}{8} \times \frac{9}{3} = \frac{1}{4} \times 3 = \frac{3}{4}$$

Example 1. $\frac{4}{5} \times \frac{2}{3} = \frac{4 \times 2}{5 \times 3} = \frac{8}{15}$

Example 2. $\frac{5}{8} \times \frac{16}{15} = \frac{5 \times 16}{8 \times 15} = \frac{80}{120} = \frac{80 \div 40}{120 \div 40} = \frac{2}{3}$

Alternative solution: $\frac{5}{8} \times \frac{16}{15} = \frac{5 \times 16}{8 \times 15} = \frac{5}{15} \times \frac{16}{8} = \frac{1}{3} \times 2 = \frac{2}{3}$

1) $\frac{3}{4} \times \frac{5}{7} =$

2) $\frac{1}{2} \times \frac{4}{7} =$

3) $\frac{4}{3} \times \frac{5}{8} =$

4) $\frac{3}{4} \times \frac{3}{4} =$

5) $\frac{4}{3} \times \frac{3}{4} =$

6) $\frac{9}{5} \times \frac{10}{3} =$

7) $\frac{1}{2} \times \frac{1}{3} =$

8) $\frac{7}{12} \times \frac{8}{3} =$

9) $\frac{15}{16} \times \frac{4}{5} =$

10) $\frac{3}{5} \times \frac{25}{12} =$

3.5 Reciprocals

In Sec. 1.5, we learned that a power of -1 makes a reciprocal. Recall that the reciprocal of a whole number is one over the number. For example, $3^{-1} = \frac{1}{3}$. The reciprocal of a fraction is found by swapping the numerator and denominator. For example, the reciprocal of $\frac{2}{5}$ is equal to $\left(\frac{2}{5}\right)^{-1} = \frac{5}{2}$.

Example 1. $\left(\frac{2}{3}\right)^{-1} = \frac{3}{2}$

Example 2. $5^{-1} = \frac{1}{5}$

Example 3. $\left(\frac{15}{11}\right)^{-1} = \frac{11}{15}$

Example 4. $\left(\frac{1}{6}\right)^{-1} = \frac{6}{1} = 6$

1) $\left(\frac{4}{3}\right)^{-1} =$

2) $\left(\frac{5}{7}\right)^{-1} =$

3) $\left(\frac{3}{5}\right)^{-1} =$

4) $\left(\frac{9}{10}\right)^{-1} =$

5) $\left(\frac{8}{7}\right)^{-1} =$

6) $\left(\frac{6}{11}\right)^{-1} =$

7) $\left(\frac{1}{8}\right)^{-1} =$

8) $\left(\frac{6}{7}\right)^{-1} =$

9) $\left(\frac{4}{5}\right)^{-1} =$

10) $\left(\frac{12}{7}\right)^{-1} =$

11) $\left(\frac{10}{9}\right)^{-1} =$

12) $\left(\frac{5}{6}\right)^{-1} =$

13) $\left(\frac{9}{13}\right)^{-1} =$

14) $9^{-1} =$

15) $\left(\frac{14}{11}\right)^{-1} =$

16) $\left(\frac{15}{8}\right)^{-1} =$

17) $\left(\frac{16}{13}\right)^{-1} =$

18) $\left(\frac{11}{12}\right)^{-1} =$

3.6 Dividing Fractions

Dividing by a fraction is equivalent to multiplying by its reciprocal. For example:

$$\frac{5}{4} \div \frac{2}{3} = \frac{5}{4} \times \left(\frac{2}{3}\right)^{-1} = \frac{5}{4} \times \frac{3}{2} = \frac{5 \times 3}{4 \times 2} = \frac{15}{8}$$

Note that the reciprocal of $\frac{2}{3}$ is equal to $\frac{3}{2}$.

Example 1. $\frac{8}{9} \div \frac{1}{7} = \frac{8}{9} \times \frac{7}{1} = \frac{8\times7}{9\times1} = \frac{56}{9}$

Example 2. $\frac{9}{4} \div \frac{3}{8} = \frac{9}{4} \times \frac{8}{3} = \frac{9\times8}{4\times3} = \frac{72}{12} = \frac{72\div12}{12\div12} = \frac{6}{1} = 6$

Alternative solution: $\frac{9}{4} \div \frac{3}{8} = \frac{9}{4} \times \frac{8}{3} = \frac{9\times8}{4\times3} = \frac{9}{3} \times \frac{8}{4} = 3 \times 2 = 6$

1) $\frac{3}{5} \div \frac{4}{7} =$

2) $\frac{5}{4} \div \frac{1}{2} =$

3) $\frac{8}{9} \div \frac{2}{3} =$

4) $\frac{1}{3} \div \frac{1}{4} =$

5) $\dfrac{2}{5} \div \dfrac{2}{5} =$

6) $\dfrac{5}{2} \div \dfrac{2}{5} =$

7) $\dfrac{3}{7} \div \dfrac{6}{5} =$

8) $\dfrac{5}{12} \div \dfrac{3}{4} =$

9) $\dfrac{21}{20} \div \dfrac{7}{5} =$

10) $\dfrac{11}{6} \div \dfrac{2}{3} =$

11) $\dfrac{9}{10} \div 3 =$

3.7 Powers of Fractions

If a fraction is raised to a power, apply the power to both the numerator and denominator.

For example, $\left(\frac{5}{6}\right)^4 = \frac{5^4}{6^4} = \frac{5\times5\times5\times5}{6\times6\times6\times6} = \frac{25\times25}{36\times36} = \frac{625}{1296}$.

Example 1. $\left(\frac{4}{3}\right)^2 = \frac{4^2}{3^2} = \frac{4\times4}{3\times3} = \frac{16}{9}$

Example 2. $\left(\frac{3}{8}\right)^3 = \frac{3^3}{8^3} = \frac{3\times3\times3}{8\times8\times8} = \frac{9\times3}{64\times8} = \frac{27}{512}$

1) $\left(\frac{4}{7}\right)^3 =$

2) $\left(\frac{4}{3}\right)^4 =$

3) $\left(\frac{5}{6}\right)^2 =$

4) $\left(\frac{8}{9}\right)^1 =$

5) $\left(\frac{3}{2}\right)^5 =$

6) $\left(\frac{3}{5}\right)^4 =$

7) $\left(\frac{8}{7}\right)^3 =$

8) $\left(\frac{2}{7}\right)^0 =$

9) $\left(\frac{6}{5}\right)^4 =$

10) $\left(\frac{8}{9}\right)^3 =$

3.8 Negative Powers of Fractions

A fraction raised to a negative power is equivalent to its reciprocal raised to a positive power:

$$\left(\frac{7}{5}\right)^{-3} = \left(\frac{5}{7}\right)^{3} = \frac{5^3}{7^3} = \frac{5 \times 5 \times 5}{7 \times 7 \times 7} = \frac{25 \times 5}{49 \times 7} = \frac{125}{343}$$

Example 1. $\left(\frac{5}{4}\right)^{-4} = \left(\frac{4}{5}\right)^{4} = \frac{4^4}{5^4} = \frac{4\times4\times4\times4}{5\times5\times5\times5} = \frac{16\times16}{25\times25} = \frac{256}{625}$

Example 2. $\left(\frac{7}{9}\right)^{-2} = \left(\frac{9}{7}\right)^{2} = \frac{9^2}{7^2} = \frac{9\times9}{7\times7} = \frac{81}{49}$

1) $\left(\frac{5}{8}\right)^{-3} =$

2) $\left(\frac{4}{7}\right)^{-2} =$

3) $\left(\frac{6}{5}\right)^{-4} =$

4) $\left(\frac{8}{9}\right)^{-1} =$

5) $\left(\frac{6}{7}\right)^{-3} =$

6) $\left(\frac{2}{3}\right)^{-5} =$

7) $\left(\frac{5}{6}\right)^{-3} =$

8) $\left(\frac{9}{2}\right)^{-4} =$

9) $\left(\frac{7}{3}\right)^{-0} =$

10) $\left(\frac{3}{10}\right)^{-5} =$

3.9 Fractional Powers

When a number is raised to a fractional power, the denominator of the fractional power has the same effect as a root (Sec. 1.8). For example:

$$8^{2/3} = \left(\sqrt[3]{8}\right)^2 = 2^2 = 2 \times 2 = 4$$

A problem of the form $x^{m/n}$ can be expressed as:

$$x^{m/n} = \left(\sqrt[n]{x}\right)^m$$

Remember that if the root is even, there are two solutions (Sec. 1.8). For example, $\sqrt[4]{16} = \pm 2$.

Example 1. $(49)^{3/2} = \left(\sqrt{49}\right)^3 = (\pm 7)^3 = \pm 7 \times 7 \times 7 = \pm 49 \times 7 = \pm 343$

Example 2. $(32)^{4/5} = \left(\sqrt[5]{32}\right)^4 = 2^4 = 2 \times 2 \times 2 \times 2 = 4 \times 4 = 16$

Example 3. $(256)^{1/4} = \left(\sqrt[4]{256}\right)^1 = \pm 4$

Example 4. $(243)^{-3/5} = \left(\frac{1}{243}\right)^{3/5} = \left(\frac{1}{\sqrt[5]{243}}\right)^3 = \frac{1}{3^3} = \frac{1}{3 \times 3 \times 3} = \frac{1}{9 \times 3} = \frac{1}{27}$

Example 5. $\left(\frac{8}{27}\right)^{5/3} = \left(\sqrt[3]{\frac{8}{27}}\right)^5 = \left(\frac{\sqrt[3]{8}}{\sqrt[3]{27}}\right)^5 = \frac{2^5}{3^5} = \frac{2 \times 2 \times 2 \times 2 \times 2}{3 \times 3 \times 3 \times 3 \times 3} = \frac{4 \times 4 \times 2}{9 \times 9 \times 3} = \frac{16 \times 2}{81 \times 3} = \frac{32}{243}$

Example 6. $\left(\frac{25}{64}\right)^{-1/2} = \left(\frac{64}{25}\right)^{1/2} = \left(\sqrt{\frac{64}{25}}\right)^1 = \left(\frac{\sqrt{64}}{\sqrt{25}}\right)^1 = \left(\pm \frac{8}{5}\right)^1 = \pm \frac{8}{5}$

1) $27^{4/3} =$

2) $64^{7/6} =$

3) $81^{-3/4} =$

4) $4^{-1/2} =$

5) $100{,}000^{6/5} =$

6) $125^{-1/3} =$

7) $216^{2/3} =$

8) $1024^{-4/5} =$

9) $100^{-5/2} =$

10) $64^{-5/3} =$

11) $36^{3/2} =$

12) $\left(\frac{625}{256}\right)^{1/4} =$

13) $\left(\frac{49}{81}\right)^{-3/2} =$

14) $\left(\frac{1}{343}\right)^{-1/3} =$

15) $\left(\frac{32}{243}\right)^{4/5} =$

16) $\left(\frac{729}{1,000,000}\right)^{-5/6} =$

3.10 Rationalize the Denominator

In Sec. 1.11, we observed that a square root multiplied by itself removes the radical sign. For example, $\sqrt{4} \times \sqrt{4} = 4$ (since $\sqrt{4} = 2$ and $2 \times 2 = 4$). This is true for any number inside the radical (provided that the same number is repeated), such as $\sqrt{3} \times \sqrt{3} = 3$.

If a fraction has a square root in its denominator, the fraction can be rationalized as follows: Multiply both the numerator and the denominator by the square root in the denominator. For example:

$$\frac{1}{\sqrt{5}} = \frac{1 \times \sqrt{5}}{\sqrt{5} \times \sqrt{5}} = \frac{\sqrt{5}}{5}$$

In the last step, we used the fact that $\sqrt{5} \times \sqrt{5} = 5$.

Example 1. $\frac{1}{\sqrt{11}} = \frac{1 \times \sqrt{11}}{\sqrt{11} \times \sqrt{11}} = \frac{\sqrt{11}}{11}$

Example 2. $\frac{34}{\sqrt{17}} = \frac{34 \times \sqrt{17}}{\sqrt{17} \times \sqrt{17}} = \frac{34 \times \sqrt{17}}{17} = \frac{34 \times \sqrt{17} \div 17}{17 \div 17} = \frac{2 \times \sqrt{17}}{1} = 2 \times \sqrt{17}$

Note: We reduced the fraction using the technique from Sec. 3.1.

1) $\frac{1}{\sqrt{3}} =$

2) $\frac{6}{\sqrt{2}} =$

3) $\frac{7}{\sqrt{7}} =$

4) $\frac{50}{\sqrt{10}} =$

5) $\frac{1}{2 \times \sqrt{6}} =$

4 DECIMALS

A decimal expresses a number as a fraction with a denominator that is a power of ten, such as 10, 100, or 1000. For example, 0.3 means $\frac{3}{10}$ and 0.79 means $\frac{79}{100}$. You can figure out what the denominator equals by looking at the place value of the final decimal position. We will use the number 1,234.567 to illustrate **place value**:

- The 1 is in the thousands place, whereas the 7 is in the **thousandths** place.
- The 2 is in the hundreds place, whereas the 6 is in the **hundredths** place.
- The 3 is in the tens place, whereas the 5 is in the **tenths** place.
- The 4 is in the units place.

Note that the "th" at the end makes a very significant difference. Read carefully.

Leading zeroes (that come after the **decimal point** and before nonzero digits) are important. For example, 0.002 is much different from 0.2 or 0.000002. In contrast, **trailing zeroes** don't affect the numerical value of a number. For example, 3.200, 3.20, 3.2 and 3.2000000 all equal the exact same value.

A nice thing about working with decimals is that you can add them without having to find a common denominator. For example, $3.42 + 1.5 = 4.92$ (we'll explore this in Sec. 4.7).

Multiplying decimals works very much like multiplying integers, except that you must take care with the decimal position (as we'll explore in Sec. 4.8).

You could divide decimals using long division, but we'll learn an alternative in Sec. 4.9.

Any fraction can be converted into a decimal, and vice-versa, as we'll see in Sec.'s 4.4-4.6. Some fractions turn into **repeating decimals** (as we'll learn in Sec. 4.6). For example, the seemingly simple fraction $\frac{1}{3}$ becomes 0.3333333... (with the 3 repeating forever) in decimal form. Since it would take forever to write the complete number, we place a bar over a single three ($0.\bar{3}$) and let the bar indicate that the digit repeats. Sometimes multiple digits repeat. For example, $\frac{2}{11}$ becomes $0.\overline{18}$ (which means 0.1818181818...) in decimal form.

Decimals are commonly used in finance, science, and engineering. Why? Decimals are easy to work with because they involve powers of ten. (As we mentioned earlier, you don't need to find a common denominator in order to add or subtract them.) Scientists employ a special notation called **scientific notation**, which expresses numbers in the form 5.28×10^3 (Sec. 4.3).

4.1 Place Value

The first digit after the decimal point is in the tenths place, the second digit after the decimal point is in the hundredths place, the third digit after the decimal point is in the thousandths place, the fourth digit after the decimal point is in the ten thousandths place, etc. For example, in 9.81 the 8 is in the tenths place and the 1 is in the hundredths place.

Example 1. In 1.04, the 0 is in the tenths place and the 4 is in the hundredths place.

Example 2. In 23.5418, the 5 is in the tenths place, the 4 is in the hundredths place, the 1 is in the thousandths place, and the 8 is in the ten thousandths place.

Directions: Indicate the place value of each digit to the right of the decimal point.

1) 0.24

2) 1.732

3) 9.7531

4) 3.14159

4.2 Powers of Ten

A base of 10 raised to an exponent is called a power of ten. Examples include 10^5 and 10^{-2}. When the power is a positive integer, the answer has one followed by the same number of zeroes as the power. For example:

$$10^5 = 10 \times 10 \times 10 \times 10 \times 10 = 100{,}000 \quad \text{(5 zeroes after)}$$

When the power is a negative integer, the answer has one less zero than the absolute value of the power between the decimal point and the one. For example:

$$10^{-5} = \frac{1}{10^5} = \frac{1}{10 \times 10 \times 10 \times 10 \times 10} = \frac{1}{100{,}000} = 0.00001 \text{ (4 zeroes in between)}$$

Recall from Sec. 1.3 that $10^0 = 1$.

Example 1. $10^3 = 10 \times 10 \times 10 = 1000$ (3 zeroes after)

Example 2. $10^{-3} = \frac{1}{10^3} = \frac{1}{10 \times 10 \times 10} = \frac{1}{1000} = 0.001$ (2 zeroes in between)

1) $10^2 =$

2) $10^{-4} =$

3) $10^7 =$

4) $10^{-2} =$

5) $10^4 =$

6) $10^{-1} =$

7) $10^6 =$

8) $10^1 =$

9) $10^{-8} =$

10) $10^0 =$

11) $10^{-6} =$

12) $10^{-9} =$

4.3 Scientific Notation

Scientific notation uses a power of ten in order to position a decimal point immediately after the first digit of a number. Examples of scientific notation include 4.23×10^3 and 1.2×10^{-5}. Any number can be expressed in scientific notation by using a power of 10 to effectively move the decimal point:

- Multiply by a positive power of 10 in order to move the decimal point to the left the same number of places as the power. For example, $537.24 = 5.3724 \times 10^2$.
- Multiply by a negative power of 10 in order to move the decimal point to the right the same number of places as the power. For example, $0.081792 = 8.1792 \times 10^{-2}$.

Example 1. $54{,}200 = 5.4200 \times 10^4 = 5.42 \times 10^4$ (moved left 4 places)

Notes: We can remove trailing zeroes if they come after a decimal point. For an integer like 54,200, the decimal point comes to its right. For example, we may write $54{,}200.0 = 54{,}200$.

Example 2. $915.2 = 9.152 \times 10^2$ (moved left 2 places)

Example 3. $0.00057 = 5.7 \times 10^{-4}$ (moved right 4 places)

Directions: Express each number using scientific notation (as described above).

1) 7295

2) 0.000168

3) 0.003876

4) 638

5) 989.898

6) 0.00000687

7) 0.044

8) 8479.2

9) 412,567.43

10) 0.0000888

11) 0.21

12) 6,642,155

13) 615,842.392

14) 0.00000001

15) 0.0254123

16) 99,999.99

4.4 Converting Decimals to Fractions

To convert a decimal to a fraction, remove the decimal point and divide by the power of 10 corresponding to the final decimal position. For example, $0.3 = \frac{3}{10}$, $0.87 = \frac{87}{100}$, and $0.143 = \frac{143}{1000}$. Sometimes the answer can be reduced using the technique from Sec. 3.1. For example, $0.24 = \frac{24}{100} = \frac{24 \div 4}{100 \div 4} = \frac{6}{25}$.

Example 1. $0.9 = \frac{9}{10}$

Example 2. $0.25 = \frac{25}{100} = \frac{25 \div 25}{100 \div 25} = \frac{1}{4}$

Example 3. $2.4 = \frac{24}{10} = \frac{24 \div 2}{10 \div 2} = \frac{12}{5}$

Example 4. $0.571 = \frac{571}{1000}$

Directions: Convert each decimal to a reduced fraction.

1) $0.6 =$

2) $0.49 =$

3) $1.5 =$

4) $0.75 =$

5) $0.36 =$

6) $0.217 =$

7) $0.375 =$

8) $1.23 =$

9) $3.2 =$

10) $0.05 =$

11) $0.5 =$

12) $1.25 =$

13) $1.64 =$

14) $0.004 =$

4.5 Converting Fractions to Decimals

To convert a fraction to a decimal, multiply both the numerator and the denominator by the number needed in order to make a denominator equal to 10, 100, 1000, or other power of 10. For example, $\frac{9}{20} = \frac{9 \times 5}{20 \times 5} = \frac{45}{100}$. The denominator then indicates the final decimal position. For example, $\frac{45}{100} = 0.45$ because that puts the 5 in the hundredth's place (for the $\frac{1}{100}$).

Example 1. $\frac{1}{2} = \frac{1 \times 5}{2 \times 5} = \frac{5}{10}$

Example 2. $\frac{7}{4} = \frac{7 \times 25}{4 \times 25} = \frac{175}{100} = 1.75$

Example 3. $\frac{7}{8} = \frac{7 \times 125}{8 \times 125} = \frac{875}{1000} = 0.875$

Example 4. $\frac{17}{50} = \frac{17 \times 2}{50 \times 2} = \frac{34}{100} = 0.34$

Directions: Convert each fraction to a decimal.

1) $\frac{3}{4} =$

2) $\frac{3}{2} =$

3) $\frac{2}{5} =$

4) $\frac{5}{8} =$

5) $\frac{17}{20} =$

6) $\frac{9}{40} =$

7) $\frac{4}{25} =$

8) $\frac{8}{5} =$

9) $\frac{8}{125} =$

10) $\frac{301}{250} =$

11) $\frac{7}{10} =$

12) $\frac{41}{200} =$

4.6 Repeating Decimals

The decimal system is based on powers of 10. The factors of 10 are 2 and 5. That is, $2 \times 5 = 10$. Because of this, if the denominator of a fraction can't be factored exclusively in terms of 2's and 5's, the fraction is equivalent to a repeating decimal. For example, $\frac{1}{3} = 0.33333333...$ with the 3 repeating forever. We place a bar over a group of numbers to indicate repeating decimals. For example, $0.\overline{3} = 0.33333333...$ and $0.\overline{27} = 0.27272727...$ To convert a fraction to a repeating decimal:

- If possible, multiply the numerator and denominator both by the value needed to make a denominator of 9, 99, 999, etc. If you can do this (and the numerator is less than the denominator), you get one repeating digit for each 9. For example, $\frac{1}{3} = \frac{1\times3}{3\times3} = \frac{3}{9} = 0.\overline{3}$ and $\frac{4}{11} = \frac{4\times9}{11\times9} = \frac{36}{99} = 0.\overline{36}$. If there are more 9's in the denominator than there are digits in the numerator, there will be a repeating zero. For example, $\frac{4}{99} = 0.\overline{04}$.

- If the denominator ends in 0, write the fraction as $\frac{1}{10}$ (or $\frac{1}{100}$, etc.) times a fraction like those mentioned in the first bullet point. For example, $\frac{1}{30} = \frac{1}{10} \times \frac{1}{3} = \frac{1}{10} \times \frac{1\times3}{3\times3} = \frac{1}{10} \times \frac{3}{9} = \frac{1}{10} \times 0.\overline{3} = 0.0\overline{3}$. (This zero doesn't repeat, unlike the second zero in $0.\overline{04}$.)

- If the numerator is greater than the denominator, first convert the fraction to a mixed number (Sec. 3.2). Write the mixed number as the whole number plus a fraction, and then convert the fraction. For example, $\frac{11}{9} = 1\frac{2}{9} = 1 + \frac{2}{9} = 1 + 0.\overline{2} = 1.\overline{2}$.

Example 1. $\frac{25}{33} = \frac{25\times3}{33\times3} = \frac{75}{99} = 0.\overline{75}$

Example 2. $\frac{4}{111} = \frac{4\times9}{111\times9} = \frac{36}{999} = 0.\overline{036}$

Example 3. $\frac{7}{90} = \frac{1}{10} \times \frac{7}{9} = \frac{1}{10} \times 0.\overline{7} = 0.0\overline{7}$

Example 4. $\frac{4}{3} = 1\frac{1}{3} = 1 + \frac{1}{3} = 1 + 0.\overline{3} = 1.\overline{3}$

1) $\frac{2}{3} =$

2) $\frac{8}{11} =$

3) $\frac{16}{333} =$

4) $\frac{5}{9} =$

5) $\frac{1}{90} =$

6) $\frac{5}{3} =$

7) $\frac{43}{99} =$

8) $\frac{31}{111} =$

9) $\frac{7}{30} =$

10) $\frac{14}{9} =$

11) $\frac{2}{45} =$

12) $\frac{4}{15} =$

13) $\frac{28}{11} =$

14) $\frac{1}{900} =$

15) $\frac{5}{6} =$

16) $\frac{19}{180} =$

17) $\frac{301}{300} =$

18) $\frac{11}{12} =$

19) $\frac{7}{60} =$

20) $\frac{8}{27} =$

4.7 Adding and Subtracting Decimals

If you line two decimals up at their decimal points, you can add or subtract them just like you would add or subtract integers. Preserve the position of the decimal point in your answer. It is sometimes helpful to add one or more trailing zeroes (after the last decimal digit). If one of the numbers is an integer, you can add .0 to it (for example, $3 = 3.0 = 3.00$ etc.).

Example 1.

$$\begin{array}{r} 17.9 \\ +18.4 \\ \hline 36.3 \end{array}$$

Example 2.

$$\begin{array}{r} 0.40 \\ -0.24 \\ \hline 0.16 \end{array}$$

Example 3.

$$\begin{array}{r} 3.00 \\ +1.73 \\ \hline 4.73 \end{array}$$

Example 4.

$$\begin{array}{r} 1.3 \\ -0.4 \\ \hline 0.9 \end{array}$$

First line up the decimals at their decimal points.

1) $4.8 + 6.5 =$

2) $11.4 - 3.8 =$

3) $0.8 - 0.54 =$

4) $2.74 + 4.36 =$

5) $0.35 + 0.47 =$

6) $0.072 - 0.05 =$

7) $4.2 - 1.34 =$

8) $4.58 + 0.892 =$

9) $0.0024 + 0.0073 =$

10) $4 - 1.83 =$

11) $21.4 - 7.9 =$

12) $0.164 + 0.38 =$

4.8 Multiplying Decimals

To multiply two decimals, follow these steps:

- First ignore the decimal point and multiply the numbers.
- Now insert a decimal point such that the answer has the same number of decimals as the original decimals combined together.
- If the answer has a trailing zero, like 3.240, you may remove it: $3.240 = 3.24$.

Example 1. $0.4 \times 0.12 = 0.048$ (since $4 \times 12 = 48$ and there are 3 decimal places)

Example 2. $2 \times 0.0005 = 0.0010 = 0.001$ (since $2 \times 5 = 10$ and there are 4 decimal places)

Example 3. $0.03 \times 0.05 = 0.0015$ (since $3 \times 5 = 15$ and there are 4 decimal places)

1) $0.4 \times 0.8 =$

2) $0.03 \times 0.2 =$

3) $0.6 \times 0.25 =$

4) $0.11 \times 0.12 =$

5) $1.5 \times 2.4 =$

6) $7 \times 3.5 =$

7) $0.06 \times 0.05 =$

8) $0.007 \times 0.09 =$

9) $0.004 \times 0.003 =$

10) $1.6 \times 0.06 =$

11) $14.5 \times 0.08 =$

12) $0.05 \times 0.004 =$

4.9 Dividing Decimals

One way to divide two decimals is to follow these steps:

- Express the division as a fraction. For example, $0.18 \div 0.72 = \frac{0.18}{0.72}$.

- Multiply the numerator and denominator both by the **same** power of 10 in order to remove all of the decimal points. For example, $\frac{0.18}{0.72} = \frac{0.18 \times 100}{0.72 \times 100} = \frac{18}{72}$.

- Convert the fraction to a decimal (Sec. 4.5). If the fraction can be reduced (Sec. 3.1), it may help to reduce the fraction first. For example, $\frac{18}{72} = \frac{18 \div 18}{72 \div 18} = \frac{1}{4} = \frac{1 \times 25}{4 \times 25} = \frac{25}{100} = 0.25$.

Remember to multiply or divide the numerator and denominator by the **same** value.

Example 1. $3.2 \div 4 = \frac{3.2}{4} = \frac{3.2 \times 10}{4 \times 10} = \frac{32}{40} = \frac{32 \div 4}{40 \div 4} = \frac{8}{10} = 0.8$

Note: Our goal is to make a denominator of 10, 100, etc. That's the reason we divided 32 and 40 each by 4 (even though each was divisible by 8).

Example 2. $0.09 \div 0.6 = \frac{0.09}{0.6} = \frac{0.09 \times 100}{0.6 \times 100} = \frac{9}{60} = \frac{9 \div 3}{60 \div 3} = \frac{3}{20} = \frac{3 \times 5}{20 \times 5} = \frac{15}{100} = 0.15$

Example 3. $2 \div 0.004 = \frac{2}{0.004} = \frac{2 \times 1000}{0.004 \times 1000} = \frac{2000}{4} = \frac{2000 \div 4}{4 \div 4} = \frac{500}{1} = 500$

1) $0.48 \div 0.8 =$

2) $1.7 \div 5 =$

3) $0.032 \div 0.05 =$

4) $0.6 \div 0.8 =$

5) $4.2 \div 0.6 =$

6) $0.0016 \div 0.0032 =$

7) $0.08 \div 0.4 =$

8) $1.2 \div 9.6 =$

4.10 Powers of Decimals

It may help to review Sec.'s 1.3, 1.5, 1.8, 3.8, and 3.9. In this section, we will apply the same concepts to decimals. It may also help to review Sec. 4.8 (multiplying decimals).

Example 1. $0.3^2 = 0.3 \times 0.3 = 0.09$

Example 2. $\sqrt{0.25} = \pm 0.5$ since $0.5^2 = 0.5 \times 0.5 = 0.25$

Example 3. $0.4^3 = 0.4 \times 0.4 \times 0.4 = 0.16 \times 0.4 = 0.064$

Example 4. $\sqrt[3]{0.008} = 0.2$ since $0.2^3 = 0.2 \times 0.2 \times 0.2 = 0.04 \times 0.2 = 0.008$

Example 5. $0.25^{-1} = \frac{1}{0.25} = \frac{1 \times 4}{0.25 \times 4} = \frac{4}{1} = 4$

Example 6. $0.125^{2/3} = \left(\sqrt[3]{0.125}\right)^2 = 0.5^2 = 0.5 \times 0.5 = 0.25$

Example 7. $0.04^{-3/2} = \frac{1}{0.04^{3/2}} = \frac{1}{\left(\sqrt{0.04}\right)^3} = \frac{1}{(\pm 0.2)^3} = \frac{\pm 1}{0.008} = \frac{\pm 1 \times 1000}{0.008 \times 1000} = \frac{\pm 1000}{8} = \pm 125$

1) $0.9^2 =$

2) $0.3^3 =$

3) $\sqrt{0.36} =$

4) $\sqrt[3]{0.027} =$

5) $0.5^{-1} =$

6) $2.5^{-1} =$

7) $0.001^{1/3} =$

8) $0.16^{-1/2} =$

9) $0.04^{3/2} =$

10) $0.0001^{-3/4} =$

4.11 Dollars and Cents

In terms of dollars ($), the value of a penny is $0.01, the value of a nickel is $0.05, the value of a dime is $0.1, and the value of a quarter is $0.25.

Example 1. How many quarters equate to $1.75?

$$\$1.75 \div \$0.25 = \frac{1.75}{0.25} = \frac{1.75 \times 4}{0.25 \times 4} = \frac{7}{1} = 7 \div 1 = 7$$

Example 2. What is the value of 9 quarters, 3 dimes, 4 nickels, and 12 pennies in dollars?

$$9 \times \$0.25 + 3 \times \$0.1 + 4 \times \$0.05 + 12 \times \$0.01 = \$2.25 + \$0.3 + \$0.2 + \$0.12 = \$2.87$$

Example 3. A banana costs $0.75 and an apple costs $0.6. How much does it cost to buy 2 bananas and 5 apples?

$$2 \times \$0.75 + 5 \times \$0.6 = \$1.5 + \$3 = \$4.5$$

1) How many nickels equate to $3.25?

2) How many dimes equate to $7.60?

3) How many pennies equate to $1.63?

4) How many quarters equate to $8.75?

5) How many nickels equate to $6?

6) What is the value of 15 quarters, 7 dimes, 5 nickels, and 18 pennies in dollars?

7) An orange costs $0.8 and a watermelon costs $2.25. How much does it cost to buy 7 oranges and 2 watermelons?

8) A pencil costs $0.18. An eraser costs $0.42. A pen costs $1.25. How much does it cost to buy 5 pencils, 3 erasers, and 2 pens?

9) A rose costs $12.5. A violet costs $16.25. A lily costs $5.75. How much does it costs to buy 12 roses, 5 violets, and 8 lilies?

5 PERCENTS

A **percent** is a fraction of one hundred. For example, 42% is equivalent to 42 out of 100. Note that 100% = 1 because 100 out of 100 is equal to $\frac{100}{100} = 1$.

Pay attention to the different ways that the words percent and percentage are used:

- A **percent** refers to a specific amount, like 64% or 125%.
- A **percentage** refers to an unspecified amount, like "a percentage of the voters."

Compare "70 percent of the students" to "a percentage of the students." In the first case, the value of 70% was specified, whereas in the second case the exact amount is unknown.

To convert a decimal to a percent, multiply by 100%. For example, $0.27 = 0.27 \times 100\% = 27\%$. To convert a percent to a decimal, divide by 100%. For example, $83\% = \frac{83\%}{100\%} = 0.83$.

To convert a percent to a fraction, write the percent over 100% and (if applicable) reduce the fraction. For example, $64\% = \frac{64\%}{100\%} = \frac{64 \div 4}{100 \div 4} = \frac{16}{25}$. To convert a fraction to a percent, multiply the numerator and denominator both by the value needed to make a denominator of 100. The percent will then equal the numerator. For example, $\frac{2}{5} = \frac{2 \times 20}{5 \times 20} = \frac{40}{100} = 40\%$.

Many stores discount products as a percent off the regular price. For example, if the regular price of a suit is $150 and the suit is on sale for 20% off, the discount is $0.2 \times \$150 = \$30$ (since $20\% = \frac{20\%}{100\%} = 0.2$ when converted from a percent to a decimal). The sale price of the suit is then $\$150 - \$30 = \$120$.

Sales tax is also figured as a percent. For example, if a television costs $400 and there is an 8% sales tax, the sales tax comes to $0.08 \times \$400 = \$32$ (since $8\% = \frac{8\%}{100\%} = 0.08$). The total cost is then $\$400 + \$32 = \$432$.

Interest rates are typically expressed as percents. For example, if $200 is invested in a savings account that pays 3% interest, in one year the savings account will earn $0.03 \times \$200 = \$6$ (since $3\% = \frac{3\%}{100\%} = 0.03$).

Note that a percent increase or decrease is compared to 100%. For example:

- If a quantity increases by 20%, this means is it 120% of the original value (since 100% + 20% = 120%). This means to multiply the original value by 1.2.
- If a quantity decreases by 35%, this means it is 65% of the original value (since 100% − 35% = 65%). This means to multiply the original value by 0.65.

We will explore percent increase and decrease in Sec. 5.7.

5.1 Converting Decimals to Percents

To convert a decimal to a percent, multiply by 100%. It may help to review Sec. 4.8.

Example 1. $0.34 = 0.34 \times 100\% = 34\%$ **Example 2.** $1.4 = 1.4 \times 100\% = 140\%$

Note: This is equivalent to shifting the decimal point two places to the right.

Directions: Convert each decimal to a percent.

1) $0.91 =$

2) $0.03 =$

3) $2.2 =$

4) $0.362 =$

5) $0.5 =$

6) $0.005 =$

7) $1.23 =$

8) $0.048 =$

9) $0.125 =$

10) $0.2468 =$

11) $3 =$

12) $0.0841 =$

13) $0.101 =$

14) $0.74 =$

15) $0.9 =$

16) $0.025 =$

17) $1.414 =$

18) $0.0008 =$

5.2 Converting Percents to Decimals

To convert a percent to a decimal, divide by 100%. It may help to review Sec. 4.9.

Example 1. $29\% = \frac{29\%}{100\%} = 0.29$ **Example 2.** $5.4\% = \frac{5.4\%}{100\%} = 0.054$

Note: This is equivalent to shifting the decimal point two places to the left.

Directions: Convert each percent to a decimal.

1) 82% =

2) 6% =

3) 0.7% =

4) 135% =

5) 10% =

6) 100% =

7) 1% =

8) 0.1% =

9) 0.04% =

10) 32.6% =

11) 49% =

12) 700% =

13) 0.006% =

14) 1.357% =

15) 0.738% =

16) 250% =

17) 4.2% =

18) 0.0375% =

5.3 Converting Percents to Fractions

To convert a percent to a fraction, follow these steps:

- Write the percent over 100%. For example, $28\% = \frac{28}{100}$. (The % signs cancel.)
- If possible, reduce the fraction (Sec. 3.1). For example, $\frac{28}{100} = \frac{28 \div 4}{100 \div 4} = \frac{7}{25}$.

Example 1. $60\% = \frac{60}{100} = \frac{60 \div 20}{100 \div 20} = \frac{3}{5}$

Example 2. $5\% = \frac{5}{100} = \frac{5 \div 5}{100 \div 5} = \frac{1}{20}$

Directions: Convert each percent to a fraction.

1) $50\% =$

2) $8\% =$

3) $80\% =$

4) $150\% =$

5) $10\% =$

6) $75\% =$

7) $0.5\% =$

8) $36\% =$

9) $125\% =$

10) $4\% =$

11) $12.5\% =$

12) $0.2\% =$

5.4 Converting Fractions to Percents

To convert a fraction to a percent, follow these steps:

- Multiply the numerator and denominator both by the (same) value needed to make a denominator of 100. For example, $\frac{16}{25} = \frac{16 \times 4}{25 \times 4} = \frac{64}{100}$.

- When the denominator is 100, the percent is the numerator. For example, $\frac{64}{100} = 64\%$.

Example 1. $\frac{4}{5} = \frac{4 \times 20}{5 \times 20} = \frac{80}{100} = 80\%$

Example 2. $\frac{3}{10} = \frac{3 \times 10}{10 \times 10} = \frac{30}{100} = 30\%$

Directions: Convert each fraction to a percent.

1) $\frac{1}{4} =$

2) $\frac{9}{20} =$

3) $\frac{2}{5} =$

4) $\frac{5}{2} =$

5) $\frac{11}{20} =$

6) $\frac{9}{10} =$

7) $\frac{12}{25} =$

8) $\frac{7}{5} =$

9) $\frac{9}{50} =$

10) $\frac{1}{8} =$

5.5 Discounts and Sales Tax

Discounts and sales tax are often computed based on a percentage:

- First convert each percent to a decimal by dividing by 100% (Sec. 5.2).
- Multiply the regular price of the item by the decimal value for the discount. This is the amount of the discount.
- Subtract the amount of the discount from the regular price. This is the sale price.
- If there is more than one item, add the sale prices together to get the subtotal.
- Multiply the subtotal by the decimal value for the tax. This is the amount of the tax.
- Add the amount of the tax to the subtotal. This is the total cost.

Example 1. A backpack has a regular price of $20 in a state that does not charge a sales tax. A customer has a coupon for 35% off. What is the total cost of the purchase?

- Convert the discount rate to a decimal: $\frac{35\%}{100\%} = 0.35$.
- Multiply $20 by 0.35 to get the amount of the discount: $0.35 \times \$20 = \$7$.
- Subtract $7 from $20 to get the sale price: $\$20 - \$7 = \$13$.
- Since there is no tax, $13 is the total cost.

Example 2. A gallon of milk is priced at $4.50. It is not on sale. There is a sales tax of 4% on groceries. What is the total cost of the purchase?

- Convert the tax rate to a decimal: $\frac{4\%}{100\%} = 0.04$.
- Multiply $4.5 by 0.04 to get the amount of the tax: $0.04 \times \$4.5 = \$0.18$.
- Add $0.18 to $4.5 to get the total cost: $\$4.5 + \$0.18 = \$4.68$.

Example 3. A magnetic dry erase board has a regular price of $150. It is on sale for 20% off. There is a sales tax of 8%. What is the total cost of the purchase?

- Convert the discount and tax rates to decimals: $\frac{20\%}{100\%} = 0.2$ and $\frac{8\%}{100\%} = 0.08$.
- Multiply $150 by 0.2 to get the amount of the discount: $0.2 \times \$150 = \$30$.
- Subtract $30 from $150 to get the sale price: $\$150 - \$30 = \$120$.
- Multiply $120 by 0.08 to get the amount of the tax: $0.08 \times \$120 = \$9.6$.
- Add $9.6 to $120 to get the total cost: $\$120 + \$9.6 = \$129.6$.

1) A desk chair has a regular price of $80. The store is having a 15% off sale for office furniture. The customer is exempt from paying sales tax. What is the total cost of the purchase?

2) A pack of chalk is priced at $5. It is not on sale. There is a sales tax of 9%. What is the total cost of the purchase?

3) A cuckoo clock has a regular price of $24. It is on sale for 25% off. There is a sales tax of 10%. What is the total cost of the purchase?

5.6 Simple Interest

Simple interest calculations are based on a percentage:
- The amount of money invested or borrowed is called the **principal**.
- Convert the interest rate to a decimal by dividing by 100% (Sec. 5.2).
- Multiply the principal by the decimal value for the interest rate. This is the amount of the interest that is earned (on an investment) or that must be repaid (on a loan).
- After one period (a year, for example), the new balance equals the original investment plus the amount of the interest.

Example 1. The chess club invests $180 in a savings account that earns interest at a rate of 3% annually. What will the balance be after one year?
- Convert the interest rate to a decimal: $\frac{3\%}{100\%} = 0.03$.
- Multiply $180 by 0.03 to get the amount of the interest: $0.03 \times \$180 = \$5.4$.
- Add $5.4 to $180 to get the new balance: $\$180 + \$5.4 = \$185.4$.

Example 2. A man borrows $500 to buy a new riding lawnmower at an interest rate of 12% annually. If he pays the balance off after one year, what total amount was paid?
- Convert the interest rate to a decimal: $\frac{12\%}{100\%} = 0.12$.
- Multiply $500 by 0.12 to get the amount of the interest: $0.12 \times \$500 = \$60$.
- Add $60 to $500 to get the total amount paid: $\$500 + \$60 = \$560$.

1) A woman invests $3000 in stocks. After one year, she receives a dividend of 6%. What is the balance after one year?

2) A business takes out a loan for $1500 at an interest rate of 7% annually. If the loan is repaid after one year, what total amount is paid?

5.7 Percent Increase and Decrease

If a quantity experiences a percent increase or a percent decrease, follow these steps to find the total amount after the change:

- Divide the percent increase or decrease by 100% in order to express it as a decimal (Sec. 5.2). For example, $\frac{30\%}{100\%} = 0.3$.
- For a percent increase, add the decimal to 1 (for example, $1 + 0.3 = 1.3$). For a percent decrease, subtract the decimal from 1 (for example, $1 - 0.3 = 0.7$). (Why? It's because 100% represents the original amount and 100% = 1 in decimal form.)
- Multiply the original amount by the previous answer. This is the total amount after the change.

Example 1. The price of gasoline was \$2.40 per gallon before the price went up by 10%. What was the price per gallon after the change?

- Convert the percent to a decimal: $\frac{10\%}{100\%} = 0.1$.
- Add 0.1 to 1 since this problem involves a percent increase: $1 + 0.1 = 1.1$.
- Multiply \$2.4 by 1.1 to get the price per gallon after the change: $1.1 \times \$2.4 = \$2.64$.

Alternate solution: Add the percent change, which equals $0.1 \times \$2.40 = \$0.24$, to the original amount: $\$2.4 + \$0.24 = \$2.64$.

Example 2. A company sold 720 laptops in August. Sales dropped 5% in September. How many laptops did the company sell in September?

- Convert the percent to a decimal: $\frac{5\%}{100\%} = 0.05$.
- Subtract 0.05 from 1 since this problem involves a percent decrease: $1 - 0.05 = 0.95$.
- Multiply 720 by 0.95 to get the sales for September: $0.95 \times 720 = 684$.

Alternate solution: Subtract the percent change, which equals $0.05 \times 720 = 36$, from the original amount: $720 - 36 = 684$.

1) A stock was valued at \$42 per share before the value rose by 15%. What was the value of the stock after the change?

2) The enrollment at a college was 3600 students in 2018. The enrollment was reduced by 8% in 2019. How many students were enrolled at the college in 2019?

3) A student earned a score of 80% on her first algebra exam. On her second exam, her score was 10% higher. What was her grade on the second exam?

4) Lisa and Paul each invested $600. Lisa's investment resulted in a 9% gain. Paul's investment resulted in a 4% loss. After the change, how much more money did Lisa have in her investment compared to Paul?

6 WORKING WITH EXPRESSIONS

Consider this problem: When the square of a number is reduced by 13, the result is 36. This and many other problems can be solved logically and systematically by applying algebra. In algebra, letters like x and y are used to represent unknown quantities. For this problem, we could use x to represent the unknown number. The square of the unknown number would then be x^2. In terms of algebra, the problem states that $x^2 - 13 = 36$. In a course on algebra, you would learn how to solve such an equation to determine that $x = \pm 7$. However, this is a prealgebra book (not an algebra book). This chapter will introduce you to variables and how to work with them, and Chapter 7 will show you how to solve the most basic equations. This will help prepare you for an algebra course (which would go much more into depth on these topics).

Unknowns like x and y are called **variables**. These letters are called variables because their numerical values turn out to be different values in different problems. For example, although x may turn out to equal 5 in one problem, it may turn out to be 3 in another problem, and -7 in yet another problem. There are an infinite number of different problems which involve x, so it is only mere coincidence when x happens to have the same value in more than one problem. That's why we call it a variable.

A fixed value is called a **constant**. A number like 8 (or any other real number) is obviously a constant because 8 is always the same value. A formula sometimes uses a letter for a constant. For example, if a rock is dropped from various heights and the time of descent is measured, the formula $H = \frac{1}{2}gt^2$ relates the height (H) to the time (t). In this case, H and t are variables because their values are different for different heights, whereas g (gravitational acceleration) is considered a constant because it always equals 9.8 m/s² near earth's surface.

An **expression** doesn't include an equal sign (=) or inequality (like < or >). In contrast, an **equation** includes an equal sign. This chapter will focus on expressions like $3x^2$ or $4x - 2$. Chapter 7 will focus on equations like $2x - 3 = 5$. We may **simplify** an expression, whereas we may **solve** an equation. For example, $7x + 4x$ simplifies to $11x$ (we'll explore this in Sec. 6.2), whereas we can solve $2x - 3 = 5$ to determine that $x = 4$ (we'll explore this in Chapter 7). The **terms** of an expression or equation are separated by + signs, − signs, or = signs. For example, the equation $8x + 2 = 6x + 4$ has four terms: $8x$, 2, $6x$, and 4. A number multiplying a variable is called a **coefficient**. For example, in $3x^2$ and $5x$, 3 and 5 are the coefficients. Study the definitions of these vocabulary words. Authors and instructors use these terms in order to discuss algebra. The instructions will make more sense to you if you understand them.

6.1 Operations with Variables

In algebra, we **don't** use the standard times symbol (×) to represent multiplication. Why not? Since we often use x to represent a variable, it would be really easy to confuse × with x. What should you do instead?

- When multiplication involves a variable, just put the variable next to what it multiplies. For example, $2x$ means "2 times x" and $3xy^2$ means "3 times x times the square of y."
- When multiplication involves two numbers, use parentheses or use a middle dot. For example, $(2)(3) = 2(3) = (2)3 = 2{\cdot}3$ all mean "2 times 3."

In algebra, we usually make a fraction to represent division. For example, $\frac{x}{2}$ means "x divided by 2" and $\frac{4-x}{3}$ means "subtract x from 4 and then divide by 3." We usually **don't** use the ÷ or / symbols to represent division in algebra.

In algebra, we usually **don't** work with percentages. It's customary to convert percentages to decimals instead. For example, we would write $0.7x$ to represent "70% of x."

Example 1. $9x^2$ means 9 times the square of x **Example 2.** $5 + x$ means 5 more than x

Example 3. $\frac{x}{8}$ means x divided by 8   **Example 4.** $0.2x$ means 20% of x

Directions: Express each operation in words. (There may be more than one correct answer.)

1) $7 - x$

2) $\frac{x}{5}$

3) $4x + 1$

4) $0.75x$

5) $\frac{2}{x}$

6) $x + yz$

7) $\sqrt{3x}$

6.2 Combine Like Terms

If you have 3 apples and get 2 more apples, you could add these together to determine that you now have 3 apples + 2 apples = 5 apples. Similarly, in algebra we can add $3x + 2x = 5x$. However, if you have 3 apples and 2 oranges, it wouldn't make sense to add them together and claim that you have 5 "apples," since 2 of them aren't apples (they are oranges). Similarly, you **can't** add $3x$ and $2x^2$ together to get 5 of x because x and x^2 are unlike.

You can add or subtract **like terms** together. Like terms have the same power of the same variable. Examples of like terms include:

- $x, 2x, 3x, -8x, 1.5x,$ and $\frac{x}{4}$ (these all have a number times x)
- $5, 7, -12, 4.2,$ and $\frac{2}{3}$ (these are all constants with no variable)
- $x^2, 6x^2, -9x^2,$ and $\frac{x^2}{2}$ (these all have a number times x^2)

You can combine like terms by adding or subtracting the **coefficients**. (The coefficients are the numbers that multiply the variables. For example, the coefficient of $5x$ is 5.)

- $12x - 4x = 8x$
- $3x^2 + 6x^2 = 9x^2$

If you don't see a coefficient, the coefficient equals one. For example, x^2 is the same as $1x^2$.

- $2x + x = 2x + 1x = 3x$
- $x^2 + 3x^2 = 1x^2 + 3x^2 = 4x^2$

You **can't** combine **unlike terms** the way that you can add or subtract the coefficients of like terms. Examples of unlike terms include:

- $2x$ and 3 (since one has x, but the other does not)
- $4x^2$ and $7x$ (since one has x^2, but the other has x)
- $5x^3$ and $2x^2$ (since one has x^3, but the other has x^2)

Example 1. $4x + 3 + 2x + 7 = (4x + 2x) + (3 + 7) = 6x + 10$

Example 2. $8 + 9x - 2x - 5 = (9x - 2x) + (8 - 5) = 7x + 3$

Example 3. $x^2 + 2x + 3 + 4x^2 + 5x + 6 = (1x^2 + 4x^2) + (2x + 5x) + (3 + 6) = 5x^2 + 7x + 9$

Note that $x^2 = 1x^2$. When you don't see a coefficient, there is a 1 hiding there.

Example 4. $3x - 2 - x + 5x - 4 - 6 = (3x - 1x + 5x) + (-2 - 4 - 6) = 7x - 12$

Note that $-x = -1x$. When you don't see a coefficient, there is a 1 hiding there.

Directions: Simplify each expression by combining like terms together.

1) $5x + 8 - 3x + 6 =$

2) $12x + 6 + 13x + 3 =$

3) $x^2 + 8 + x^2 + 9 =$

4) $15x - 8 + 10x - 6 + 8x - 4 =$

5) $3x + x + 11x =$

6) $8x^2 + 10 - 4x^2 - 7 =$

7) $12 + 14x - 2x - 6 - 2x - 5 =$

8) $x - 3 + x - 1 + 2x - 3 =$

9) $16x^2 + 14 - 3x^2 - 12 - 2x^2 + 13 =$

10) $3x^2 + 14 + 4x^2 + 13 + 2x^2 - 11 =$

11) $8 - 17 + 4x^2 + 2x^2 + 11 + 11 =$

12) $x - 14 - x + 8x + 12 - 1 =$

13) $6x^3 + 2x^3 + x + x + 22 - 13 =$

14) $7x + 4x + 6x - 12 + 4 =$

15) $4x^2 + 9 + 6x^2 + 19x + 19 =$

16) $x^3 + 4x + 7 + 6x^3 + 9x + 12 =$

6.3 Exploring Expressions

Letters like x and y are called **variables** because their values can vary from one problem to another. For example, in one problem x might turn out to be 4 while in another problem it might turn out to be -9. Although variables are represented with letters, a variable actually represents a number. In this section, we will see this by plugging numbers in for variables.

For example, consider the expression $7x + 3x$.

- When $x = 2$, we get $7(2) + 3(2) = 14 + 6 = 20$. (Recall that parentheses or a middle dot may be used for multiplication between numbers.) Compare with $10(2) = 20$.
- When $x = 3$, we get $7(3) + 3(3) = 21 + 9 = 30$. Compare with $10(3) = 30$.
- When $x = 4$, we get $7(4) + 3(4) = 28 + 12 = 40$. Compare with $10(4) = 40$.
- For any value of x, you can verify that $7x + 3x = 10x$. This is why we were able to combine **like** terms in the previous section.

In contrast, consider the expression $3x^2 + 4x$.

- When $x = 2$, we get $3(2)^2 + 4(2) = 3(4) + 8 = 12 + 8 = 20$.
- When $x = 3$, we get $3(3)^2 + 4(3) = 3(9) + 12 = 27 + 12 = 39$.
- When $x = 4$, we get $3(4)^2 + 4(4) = 3(16) + 16 = 48 + 16 = 64$.
- When $x = 5$, we get $3(5)^2 + 4(5) = 3(25) + 20 = 75 + 20 = 95$.
- Unlike the previous set of bullet points, **unlike** terms $3x^2$ and $4x$ can't be condensed to a simpler single term that would hold for all possible values of x. (The best we could do in this case is factor, as we will see in Sec. 6.10.)

Example 1. Plug $x = 4$ into $2x^2 - 3$: $2(4)^2 - 3 = 2(16) - 3 = 32 - 3 = 29$

Example 2. Plug $x = 5$ into $8x - 2x$. Compare with $6x$.
$$8(5) - 2(5) = 40 - 10 = 30 \quad \text{agrees with} \quad 6(5) = 30$$

Example 3. Plug $x = 6$ into $\frac{x+4}{2}$: $\frac{6+4}{2} = \frac{10}{2} = 5$ (Recall that a fraction indicates division.)

Example 4. Plug $x = 3$ and $y = 4$ into $x^2 y$: $(3)^2(4) = (9)(4) = 36$

1) Plug $x = 3$ into $x^2 - 2x + 5$.

2) Plug $x = 4$ into $\frac{19-x}{3}$.

3) Plug $x = 6$ into $7x - 3x$. Compare with $4x$.

4) Plug $x = 10$ into $3x^2 + 2x^2$. Compare with $5x^2$.

5) Plug $x = 7$ into $3x - x$. Compare with $2x$.

6) Plug $x = -2$ into $3x - 4$.

7) Plug $x = -3$ into $x^3 + 2x^2 - 4x - 7$.

8) Plug $x = -5$ into $8x - x + 2x$. Compare with $9x$.

9) Plug $x = 8$ and $y = 3$ into $3x - 2y$.

10) Plug $x = 2$ and $y = 3$ into $x^3 y^2$.

11) Plug $x = 7$ and $y = 9$ into $\frac{y+6}{x-4}$.

12) Plug $x = \frac{1}{2}$ into $8x - 2x$. Compare with $6x$.

13) Plug $x = \frac{3}{2}$ into $x^2 + 2x^2$. Compare with $3x^2$.

6.4 Exploring Powers of Variables

The following formula shows how to multiply two powers of the same base.

$$x^m x^n = x^{m+n}$$

For example, with $m = 3$ and $n = 4$ we get $x^3 x^4 = x^7$. For $x = 2$, observe that:

$$2^3 = 2 \cdot 2 \cdot 2 = 8$$
$$2^4 = 2 \cdot 2 \cdot 2 \cdot 2 = 16$$
$$2^7 = 2 \cdot 2 \cdot 2 \cdot 2 \cdot 2 \cdot 2 \cdot 2 = 128$$
$$2^3 \cdot 2^4 = 8 \cdot 16 = 128 = 2^7$$

Recall that a middle dot ($\cdot$) or parentheses may be used to multiply numbers.

Why does $x^m x^n = x^{m+n}$? Consider the case $x = 10$, for which $10^m \cdot 10^n = 10^{m+n}$. This holds because 10^m tells you how many 10's to multiply together. Note that 10^m has m zeroes after the one. For example, $10^4 = 10 \cdot 10 \cdot 10 \cdot 10 = 10{,}000$ has 4 zeroes after the one. Since 10^{m+n} has $m + n$ zeroes after the one, it will equal $10^m \cdot 10^n$. More generally, note that x^m tells you have many x's are multiplying together. For example, $x^3 = xxx$. (Recall that no symbol is used to indicate multiplication between variables.) Since $m + n$ is how many x's are multiplying together in x^{m+n}, it will equal $x^m x^n$. For example, $x^2 = xx$, $x^3 = xxx$, and $x^5 = xxxxx$, which agrees with $x^2 x^3 = xxxxx = x^5 = x^{2+3}$.

Example 1. Show that $x^m x^n = x^{m+n}$ for $x = 2$, $m = 3$, and $n = 5$.

$$2^3 \cdot 2^5 = (2 \cdot 2 \cdot 2)(2 \cdot 2 \cdot 2 \cdot 2 \cdot 2) = (8)(32) = 256$$
$$2^{3+5} = 2^8 = 256$$

1) Show that $x^m x^n = x^{m+n}$ for $x = 3$, $m = 2$, and $n = 4$.

2) Show that $x^m x^n = x^{m+n}$ for $x = 10$, $m = 3$, and $n = 4$.

Note that $x^1 = x$. One way to see this is to consider that $x^2 = xx$ (since xx means x times x). Compare this to the special case of $x^m x^n = x^{m+n}$ where $m = 1$ and $n = 1$. In this case, we get $x^1 x^1 = x^2$. Since $x^2 = xx$, by comparison you should see that $x^1 = x$. Anything to the first power equals itself. For example, $7^1 = 7$.

Now consider the special case of $x^m x^n = x^{m+n}$ with $m = 0$. In this case, we get $x^0 x^n = x^n$. This can only be true in general if $x^0 = 1$. This is why $2^0 = 1$, $5^0 = 1$, and any other nonzero value raised to the power of zero equals one.

Now consider the special case of $x^m x^n = x^{m+n}$ where $n = -m$. In this case, we get $x^m x^{-m} = x^0$ since $m + (-m) = m - m = 0$. Recall that $x^0 = 1$. This means that $x^m x^{-m} = 1$. Divide both sides of $x^m x^{-m} = 1$ by x^m to see that:

$$x^{-m} = \frac{1}{x^m}$$

For example, $4^{-1} = \frac{1}{4}$ and $5^{-2} = \frac{1}{5^2} = \frac{1}{25}$. Recall that we explored negative powers in Sec. 3.8.

Example 2. Show that $x^m x^n = x^{m+n}$ for $x = 3$, $m = -1$, and $n = 2$.

$$3^{-1} \cdot 3^2 = \frac{1}{3}(3 \cdot 3) = \frac{9}{3} = 3$$
$$3^{-1+2} = 3^1 = 3$$

Example 3. Show that $x^m x^n = x^{m+n}$ for $x = 5$, $m = 2$, and $n = -2$.

$$5^2 \cdot 5^{-2} = (5 \cdot 5)\frac{1}{5 \cdot 5} = \frac{25}{25} = 1$$
$$5^{2+(-2)} = 5^0 = 1$$

3) Show that $x^m x^n = x^{m+n}$ for $x = 4$, $m = -2$, and $n = 3$.

4) Show that $x^m x^n = x^{m+n}$ for $x = 10$, $m = -3$, and $n = 3$.

The following formula shows how to divide two powers of the same base.

$$\frac{x^m}{x^k} = x^{m-k}$$

This is really the same as the rule $x^m x^n = x^{m+n}$ if you let $n = -k$, since $x^{-k} = \frac{1}{x^k}$.

Example 4. Show that $\frac{x^m}{x^k} = x^{m-k}$ for $x = 3$, $m = 4$, and $k = 2$.

$$\frac{3^4}{3^2} = \frac{3 \cdot 3 \cdot 3 \cdot 3}{3 \cdot 3} = \frac{81}{9} = 9$$
$$3^{4-2} = 3^2 = 9$$

5) Show that $\frac{x^m}{x^k} = x^{m-k}$ for $x = 2$, $m = 5$, and $k = 3$.

6) Show that $\frac{x^m}{x^k} = x^{m-k}$ for $x = 3$, $m = 6$, and $k = -2$.

Now consider a power of a power like $(x^2)^3$. What this means is to raise x^2 to the power of three: $(x^2)^3 = x^2 x^2 x^2$. This works out to $x^2 x^2 x^2 = x^{2+2+2}$. Thus, $(x^2)^3 = x^6$. A more general formula is:

$$(x^m)^n = x^{mn}$$

Example 5. Show that $(x^m)^n = x^{mn}$ for $x = 10$, $m = 2$, and $n = 3$.
$$(10^2)^3 = (100)^3 = 1,000,000$$
$$10^{2 \cdot 3} = 10^6 = 1,000,000$$

7) Show that $(x^m)^n = x^{mn}$ for $x = 2$, $m = 3$, and $n = 4$.

6.5 Working with Powers of Variables

In this section, we will apply the following rules from Sec. 6.4:

$$x^m x^n = x^{m+n} \quad , \quad x^{-m} = \frac{1}{x^m} \quad , \quad \frac{x^m}{x^k} = x^{m-k} \quad , \quad (x^m)^n = x^{mn}$$

Also note the following special cases:

$$x^0 = 1 \quad , \quad x^1 = x \quad , \quad x^2 = xx$$

Note that xx means x times x (since no symbol is used for multiplication with variables).

Example 1. $x^5 x^2 = x^{5+2} = x^7$ **Example 2.** $\dfrac{x^9}{x^6} = x^{9-6} = x^3$

Example 3. $x^4 x^{-3} = x^{4+(-3)} = x^{4-3} = x^1 = x$ **Example 4.** $\dfrac{x^3}{x^{-2}} = x^{3-(-2)} = x^{3+2} = x^5$

Example 5. $x^2 x^{-4} x^{-3} = x^{2+(-4)+(-3)} = x^{-5} = \dfrac{1}{x^5}$ **Example 6.** $\dfrac{x^{-5}}{x^{-4}} = x^{-5-(-4)} = x^{-5+4} = x^{-1} = \dfrac{1}{x}$

Example 7. $(x^3)^2 = x^{3 \cdot 2} = x^6$ **Example 8.** $\dfrac{1}{x^{-2}} = \dfrac{x^0}{x^{-2}} = x^{0-(-2)} = x^{0+2} = x^2$

Example 9. $\dfrac{x^2 x}{x^4} = \dfrac{x^2 x^1}{x^4} = \dfrac{x^3}{x^4} = x^{3-4} = x^{-1} = \dfrac{1}{x}$ **Example 10.** $\left(\dfrac{x^3}{x}\right)^4 = \left(\dfrac{x^3}{x^1}\right)^4 = (x^2)^4 = x^{2 \cdot 4} = x^8$

1) $x^3 x^2 =$ 2) $x^8 x^6 =$

3) $x^4 x =$ 4) $x^4 x^2 x^0 =$

5) $x^5 x^{-3} =$ 6) $x^{-7} x^8 =$

7) $x^{-9} x^{-8} =$ 8) $x^3 x^{-4} x =$

9) $\dfrac{x^7}{x^3} =$ 10) $\dfrac{x^{12}}{x^5} =$

11) $\dfrac{x^4}{x^4} =$ 12) $\dfrac{x^6}{x^{-6}} =$

13) $\dfrac{x^{-2}}{x^3} =$ 14) $\dfrac{x}{x^2} =$

15) $\dfrac{x^{-7}}{x^{-4}} =$ 16) $\dfrac{x^{-5}}{x^{-9}} =$

17) $x^{-3} =$

18) $\frac{1}{x^{-3}} =$

19) $\frac{x^2}{1} =$

20) $\frac{x^{-4}}{x^0} =$

21) $(x^2)^4 =$

22) $(x^5)^6 =$

23) $(x^{-2})^3 =$

24) $(x^{-3})^{-4} =$

25) $(x^{-1})^{-1} =$

26) $x^{-1}x^{-1} =$

27) $\frac{x^4 x^8}{x^2 x^3} =$

28) $\frac{x^6 x^{-3}}{x^5 x^{-4}} =$

29) $\frac{x^2 x}{x^{-3}} =$

30) $\frac{x^3 x^{-7}}{x^5} =$

31) $\frac{x^8 x^{-8}}{x^{-1}} =$

32) $\left(\frac{x^5}{x^2}\right)^3 =$

33) $\left(\frac{x^{-2}}{x^{-3}}\right)^4 =$

34) $\left(\frac{x^9 x^6}{x^8}\right)^2 =$

35) $\left(\frac{x^4}{x^5}\right)^{-1} =$

36) $\left(\frac{x^5}{x}\right)^{-2} =$

37) $\left(\frac{1}{x^3}\right)^{-4} =$

38) $\left(\frac{1}{x^{-4}}\right)^{-3} =$

39) $\left(\frac{x^3}{x^{-2}}\right)^{-5} =$

40) $\left(\frac{x^{-7}}{x^{-3}}\right)^{-1} =$

41) $\left(\frac{x^{-3}}{x}\right)^{-2} =$

42) $\left(\frac{x^{-2}}{x^{-5}}\right)^{-4} =$

43) $\left(\frac{x^8 x^7}{x^5 x^4}\right)^3 =$

44) $\left(\frac{x^5 x^{-3}}{x^2 x^{-4}}\right)^{-2} =$

Another rule for working with powers is:

$$(cx^n)^k = c^k x^{nk}$$

Here, c represents a constant like 2 or 5. For example, $(2x^3)^5 = 2^5 x^{3 \cdot 5} = 32x^{15}$.

Example 11. $(3x)^5 = 3^5 x^{1 \cdot 5} = 243x^5$

Example 12. $(7x^6)^2 = 7^2 x^{6 \cdot 2} = 49x^{12}$

Recall that $x^1 = x$.

45) $(5x)^3 =$

46) $(2x^3)^6 =$

47) $(-x^2)^5 =$

48) $(-x^7)^8 =$

49) $(8x^9)^{-2} =$

50) $(5x^0)^4 =$

51) $(-2x^{-3})^6 =$

52) $(3x^2)^{-2} =$

53) $(7x^6)^0 =$

54) $(-4x^{-5})^{-3} =$

If there are two variables inside the parentheses, the same concept still applies to each:

$$(cx^m y^n)^k = c^k x^{mk} y^{nk}$$

For example, $(6x^2 y^3)^3 = 6^3 x^{2 \cdot 3} y^{3 \cdot 3} = 216x^6 y^9$.

Example 13. $(5x^2 y)^4 = 5^4 x^{2 \cdot 4} y^{1 \cdot 4} = 625x^8 y^4$ (Note that $y^1 = y$.)

Example 14. $(4x^3 y^2)^5 = 4^5 x^{3 \cdot 5} y^{2 \cdot 5} = 1024x^{15} y^{10}$

55) $(x^7 y^6)^8 =$

56) $(9xy)^2 =$

57) $(2x^4 y^6)^5 =$

58) $(-x^6 y^3)^3 =$

59) $(6x^8 y^8)^2 =$

60) $(8x^5 y^7)^1 =$

61) $(-2x^{-7} y^{-2})^6 =$

62) $(3x^4 y^{-1})^4 =$

63) $(x^3 y^{-2})^{-1} =$

64) $(-x^5 y^{-5})^{-5} =$

6.6 Distributing with Variables

The distributive property of arithmetic is $x(y + z) = xy + xz$. You can check that this formula holds for a variety of numbers. For example, $3(4 + 5) = 3(4) + 3(5)$ because $3(4 + 5) = 3(9)$ $= 27$ and $3(4) + 3(5) = 12 + 15 = 27$. That is, both sides of $3(4 + 5) = 3(4) + 3(5)$ equal 27.

Example 1. $3(x + 2) = 3(x) + 3(2) = 3x + 6$
Example 2. $x(x - 1) = x(x) + x(-1) = x^2 - x$
Example 3. $4x(x^2 + x) = 4x(x^2) + 4x(x) = 4x^3 + 4x^2$
Example 4. $2x(-4x + 5) = 2x(-4x) + 2x(5) = -8x^2 + 10x$
Example 5. $5x^2(2x^2 + 3x - 4) = 5x^2(2x^2) + 5x^2(3x) + 5x^2(-4) = 10x^4 + 15x^3 - 20x^2$

Directions: Distribute the expression to each term in parentheses.

1) $2(x + 4) =$

2) $x(3 - x) =$

3) $3x(x - 2) =$

4) $5x(2x + 3) =$

5) $4x(-x^2 - 5) =$

6) $2x(3x^2 + 4x) =$

7) $3x^2(4x^2 - 3) =$

8) $x(x^2 - x + 1) =$

9) $3x(2x^2 + 5x - 7) =$

10) $4x^2(3x^7 - 6x^4 + 9) =$

11) $8x^4(-7x^6 - 5x^5 + 4x^4) =$

Any minus sign coming before the parentheses also gets distributed. For example:
$$-2(x + 3) = -2(x) - 2(3) = -2x - 6$$
$$-x(x - 4) = -x(x) - x(-4) = -x^2 + 4x$$
Pay close attention to how the minus sign distributes to each term in the examples. Note that two minus signs make a plus sign in multiplication. For example, $-x(-4) = 4x$.

Example 6. $-3(x + 4) = -3(x) - 3(4) = -3x - 12$

Example 7. $-(x - 1) = -(x) - (-1) = -x + 1$

Note that $-(x - 1) = -1(x - 1)$ and $-(x) = -1(x)$. A coefficient of 1 is hidden.

Example 8. $-2x(-x + 5) = -2x(-x) - 2x(5) = 2x^2 - 10x$

Example 9. $-4x(x^2 - 2x + 3) = -4x(x^2) - 4x(-2x) - 4x(3) = -4x^3 + 8x^2 - 12x$

Directions: Distribute the expression to each term in parentheses.

12) $-7(6x + 8) =$

13) $-x(4x - 5) =$

14) $-(7 - x) =$

15) $-9x(-8x^2 - 6x) =$

16) $-6x^2(-7x + 6) =$

17) $-8x(8x^2 + 6x) =$

18) $-5(x^2 - 4x + 9) =$

19) $-(2x^4 - 4x + 8) =$

20) $-x(-5x^2 + 6x - 3) =$

21) $-7x^2(8x^5 - 6x^3 + 4x) =$

22) $-9x^5(7x^8 + 5x^6 - 3x^4) =$

6.7 The FOIL Method

FOIL stands for "First, Outside, Inside, Last." This abbreviation can help you remember how to apply the distributive property to a structure like this:

$$(w + x)(y + z) = wy + wz + xy + xz$$

- The first term, wy, comes from multiplying the **first** part of each expression.
- The second term, wz, comes from multiplying the **outside** part of each expression.
- The third term, xy, comes from multiplying the **inside** part of each expression.
- The last term, xz, comes from multiplying the **last** part of each expression.

You can verify that the FOIL method works by checking a variety of numbers. For example, $(2 + 3)(4 + 6) = 2(4) + 2(6) + 3(4) + 3(6)$. The left-hand side is $(5)(10) = 50$ and the right-hand side also equals $8 + 12 + 12 + 18 = 20 + 30 = 50$.

Example 1. $(x + 3)(x - 2) = x(x) + x(-2) + 3(x) + 3(-2) = x^2 - 2x + 3x - 6 = x^2 + x - 6$

Note: In the last step, we combined like terms (Sec. 6.2): $-2x + 3x = x$.

Example 2. $(2x + 1)(3x^2 - 4) = 2x(3x^2) + 2x(-4) + 1(3x^2) + 1(-4) = 6x^3 - 8x + 3x^2 - 4$
$= 6x^3 + 3x^2 - 8x - 4$ (In the last step we reordered the terms because it is customary to express the answer in order of decreasing powers of the variable.)

Directions: Apply the FOIL method to expand each expression.

1) $(x + 4)(x + 3) =$

2) $(x - 2)(x + 5) =$

3) $(x - 1)(x - 2) =$

4) $(-x + 6)(x + 4) =$

5) $(3 - x)(8 - x) =$

6) $(-x - 5)(x + 6) =$

7) $(x^2 + x)(x + 1) =$

8) $(-x - 7)(-x - 8) =$

9) $(3x + 2)(4x + 5) =$

10) $(x - 6)(-x - 9) =$

11) $(2x^2 + 4)(3x^2 - 5) =$

12) $(5x^2 - 3)(2x + 6) =$

13) $(x^2 + 1)(x^2 - 1) =$

14) $(4x^3 + 2x)(3x^2 - 2x) =$

15) $(6x^2 - 7x)(5x - 3) =$

16) $(4x + 7)(4x + 7) =$

17) $(8x^3 + 6x)(7x^4 - 5x^2) =$

6.8 Distributing More Than Two Terms

The same concept as the FOIL method can be applied when there are more than two terms enclosed in parentheses. Each term in the first expression multiplies every term in the second expression. For example, $(v + w)(x + y + z) = vx + vy + vz + wx + wy + wz$.

Example 1. $(x - 3)(x^2 + 2x - 4) = x(x^2) + x(2x) + x(-4) - 3(x^2) - 3(2x) - 3(-4)$

$= x^3 + 2x^2 - 4x - 3x^2 - 6x + 12 = \boxed{x^3 - x^2 - 10x + 12}$

Note: In the last step, we combined like terms (Sec. 6.2): $2x^2 - 3x^2 = -x^2$ and $-4x - 6x = -10x$.

Example 2. $(x^2 + 2x - 3)(x^2 - 4x + 5) = x^2(x^2) + x^2(-4x) + x^2(5) + 2x(x^2) + 2x(-4x)$

$+2x(5) - 3(x^2) - 3(-4x) - 3(5) = x^4 - 4x^3 + 5x^2 + 2x^3 - 8x^2 + 10x - 3x^2 + 12x - 15$

$= \boxed{x^4 - 2x^3 - 6x^2 + 22x - 15}$ Note: In the last step, we combined like terms (Sec. 6.2):

$-4x^3 + 2x^3 = -2x^3, 5x^2 - 8x^2 - 3x^2 = -6x^2$, and $10x + 12x = 22x$.

Directions: Apply the distributive property to expand each expression.

1) $(x + 2)(x^2 + 4x + 3) =$

2) $(2x - 1)(x^2 - 3x + 5) =$

3) $(x^2 - 3x + 5)(x^2 + 2x - 4) =$

4) $(2x^2 - x + 4)(3x^2 + 5x - 7) =$

6.9 Expansion of Powers

If an expression of the form $(x + y)$ is raised to a power (that equals an integer of 2 or larger), multiply the expression by itself as needed. For example:

$$(x + y)^3 = (x + y)(x + y)(x + y)$$

Example 1. $(x + 5)^2 = (x + 5)(x + 5) = x(x) + x(5) + 5(x) + 5(5) = x^2 + 5x + 5x + 25$
$= x^2 + 10x + 25$ In the last step, we combined like terms (Sec. 6.2): $5x + 5x = 10x$.
Example 2. $(x - 2)^3 = (x - 2)(x - 2)(x - 2) = [x(x) + x(-2) - 2(x) - 2(-2)](x - 2)$
$= (x^2 - 2x - 2x + 4)(x - 2) = (x^2 - 4x + 4)(x - 2)$
$= x^2(x) + x^2(-2) - 4x(x) - 4x(-2) + 4(x) + 4(-2) = x^3 - 2x^2 - 4x^2 + 8x + 4x - 8$
$= x^3 - 6x^2 + 12x - 8$ Note that we used square brackets [ ] in addition to parentheses ( ) in order to help you follow the steps.

Directions: Apply the distributive property to expand each expression.

1) $(x + 4)^2 =$

2) $(3x - 5)^2 =$

3) $(x + 2)^3 =$

4) $(2x - 3)^3 =$

6.10 Factoring with Variables

Factoring is basically the distributive property in reverse. To see how, consider an example of the distributive property: $3x^2(2x - 5) = 3x^2(2x) + 3x^2(-5) = 6x^3 - 15x^2$. When this is written in reverse, we call it **factoring**:

$$6x^3 - 15x^2 = 3x^2(2x) + 3x^2(-5) = 3x^2(2x - 5)$$

Specifically, we factored $3x^2$ out of $6x^3 - 15x^2$.

To factor an expression with a single variable (like x), follow these steps:

- Identify the greatest common factor (GCF) of all the coefficients. For example, for $8x^5$ and $12x^3$, the coefficients are 8 and 12. The GCF of 8 and 12 is 4 (since 4 is the largest number for which 8 and 12 are both multiples).
- Of the given terms, what is the smallest power of the variable? For example, for $8x^5$ and $12x^3$, the smallest power is x^3 (since the alternative x^5 has a higher power).
- Write the answers to the first two steps multiplying one another times parentheses. Inside the parentheses, write the sum (or difference, if appropriate) of terms needed so that if the distributive property were applied (Sec. 6.6), the answer would equal the given expression. For example, $8x^5 + 12x^3 = 4x^3(2x^2 + 3)$ because if you distribute the $4x^3$ across the $(2x^2 + 3)$, you would get $8x^5 + 12x^3$.
- If every term is negative, also factor out a minus sign. For example
$$-3x^3 - 6x = -3x(x^2 + 2)$$
- You can check your answers easily by distributing (like we did in Sec. 6.6).

Example 1. $10x^4 + 15x = 5x(2x^3 + 3)$ **Example 2.** $4x - 12 = 4(x - 3)$

Example 3. $-24x^5 - 32x^3 = -8x^3(3x^2 + 4)$ **Example 4.** $-7x - 4 = -(7x + 4)$

Example 5. $3x^4 + 6x^3 - 3x^2 = 3x^2(x^2 + 2x - 1)$

Directions: Factor the expressions following the examples above.

1) $4x^3 + 8x =$

2) $18x^7 - 12x^3 =$

3) $5x^4 - 7x^2 =$

4) $-x^4 - x^2 =$

5) $15x + 25 =$

6) $18x^5 - 24x^2 =$

7) $-4x^2 + 6x =$

8) $48x^9 - 36x^7 =$

9) $21x^5 + 28x =$

10) $-7x^8 - 11x^5 =$

11) $12x^5 + 20x^4 - 16x^3 =$

12) $15x^4 - 9x^3 + 6x^2 =$

13) $-8x^3 + 10x^2 - 6x =$

14) $-x^8 - 3x^7 - 2x^5 =$

15) $3x^2 - 6x + 9 =$

16) $36x^9 + 27x^8 - 45x^7 =$

17) $48x^5 + 24x^4 + 60x^3 =$

18) $-45x^6 - 30x^5 + 60x^4 =$

19) $33x^3 - 22x^2 - 55x =$

20) $-32x^7 - 48x^6 - 24x^3 =$

21) $12x^6 - 18x^5 + 9x^4 =$

6.11 The Sum or Difference of Squares

It is handy to remember the following formulas because they get used frequently in algebra:
$$(w + y)(w + y) = w^2 + 2wy + y^2$$
$$(w + y)(w - y) = w^2 - y^2$$
The formulas follow from the FOIL method (Sec. 6.7). If you expand $(w + y)(w + y)$, first you get $w^2 + wy + yw + y^2$ which simplifies to $w^2 + 2wy + y^2$. If you expand $(w + y)(w - y)$, first you get $w^2 - wy + yw - y^2$ which simplifies to $w^2 - y^2$ (since $-wy + yw = 0$).

Example 1. $(x + 6)(x + 6) = x^2 + 2x(6) + 6^2 = x^2 + 12x + 36$ Let $w = x$ and $y = 6$.

Example 2. $(x + 7)(x - 7) = x^2 - 7^2 = x^2 - 49$ Let $w = x$ and $y = 7$.

Example 3. $(x - 2)(x - 2) = x^2 + 2x(-2) + (-2)^2 = x^2 - 4x + 4$ Let $w = x$ and $y = -2$.
Note: You can think of this problem as $[x + (-2)][x + (-2)]$.

Example 4. $(5x + 8)^2 = (5x + 8)(5x + 8) = (5x)^2 + 2(5x)(8) + 8^2 = 25x^2 + 80x + 64$
Let $w = 5x$ and $y = 8$.

Example 5. $(6x - 9)(6x + 9) = (6x + 9)(6x - 9) = (6x)^2 - 9^2 = 36x^2 - 81$
Let $w = 6x$ and $y = 9$. Note: We applied the commutative property of multiplication (Sec. 2.5). That is, $(w + y)(w - y) = (w - y)(w + y) = w^2 - y^2$.

Directions: Apply the sum or difference of squares formula to expand each expression.

1) $(x + 8)(x + 8) =$

2) $(x + 5)(x - 5) =$

3) $(x + 4)^2 =$

4) $(x - 9)(x + 9) =$

5) $(3x + 4)(3x + 4) =$

6) $(7x + 2)(7x - 2) =$

7) $(8x - 3)(8x - 3) =$

8) $(4x - 5)(4x + 5) =$

9) $(9x - 4)^2 =$

10) $(x^2 + 9)(x^2 - 9) =$

11) $(x^2 + 6)(x^2 + 6) =$

12) $(x^3 - 3)(x^3 + 3) =$

13) $(-x - 1)(x + 1) =$

The sum and difference of squares formulas can be used to factor expressions that have the form $w^2 + 2wy + y^2$ or the form $w^2 - y^2$. For example:

$$x^2 + 6x + 9 = x^2 + 2x3 + 3^2 = (x + 3)^2$$
$$x^2 - 16 = x^2 - 4^2 = (x + 4)(x - 4)$$

Example 6. $x^2 + 10x + 25 = x^2 + 2x5 + 5^2 = (x + 5)^2$ Let $w = x$ and $y = 5$.

Example 7. $x^2 - 36 = x^2 - 6^2 = (x + 6)(x - 6)$ Let $w = x$ and $y = 6$.

Example 8. $x^2 - 6x + 9 = x^2 + 2x(-3) + (-3)^2 = [x + (-3)]^2 = (x - 3)^2$ Let $w = x$ and $y = -3$.

Example 9. $9x^2 + 12x + 4 = (3x)^2 + 2(3x)(2) + 2^2 = (3x + 2)^2$ Let $w = 3x$ and $y = 2$.

Note: You can check the answer by distributing.

Example 10. $25x^2 - 36 = (5x)^2 - 6^2 = (5x + 6)(5x - 6)$ Let $w = 5x$ and $y = 6$.

Directions: Apply the sum or difference of squares formula to factor each expression.

14) $x^2 - 9 =$

15) $x^2 + 4x + 4 =$

16) $4x^2 - 64 =$

17) $x^2 + 18x + 81 =$

18) $x^2 - 1 =$

19) $16x^2 + 48x + 36 =$

20) $49x^2 - 100 =$

6.12 Square Roots with Variables

The following rules apply to square roots:

$$\left(\sqrt{x}\right)^2 = \sqrt{x}\sqrt{x} = x \quad , \quad \sqrt{ax} = \sqrt{a}\sqrt{x}$$

You can check that these formulas hold for a variety of numbers. For example, $\left(\sqrt{9}\right)^2 = 3^2 = 9$ agrees with $\sqrt{9}\sqrt{9} = 3 \cdot 3 = 9$, and $\sqrt{4 \cdot 9} = \sqrt{36} = 6$ agrees with $\sqrt{4}\sqrt{9} = 2 \cdot 3 = 6$. These also hold for the negative roots. For example, $(-3)(-3) = 9$. (Recall from Sec. 1.6 that $\sqrt{9} = \pm 3$.)

If a square root includes a perfect square, it is customary to factor it out of the square root. For example, $\sqrt{4x} = \sqrt{4}\sqrt{x} = 2\sqrt{x}$ and $\sqrt{3x^2} = \sqrt{3}\sqrt{x^2} = x\sqrt{3}$. It may help to review Sec. 1.10.

Example 1. $\sqrt{25x} = \sqrt{25}\sqrt{x} = 5\sqrt{x}$ **Example 2.** $\sqrt{7x^2} = \sqrt{7}\sqrt{x^2} = x\sqrt{7}$

Example 3. $\sqrt{12x} = \sqrt{4}\sqrt{3x} = 2\sqrt{3x}$ **Example 4.** $\sqrt{64x^6} = \sqrt{64}\sqrt{x^6} = 8x^3$

Note: $\sqrt{x^6} = x^3$ because $(x^3)^2 = x^6$ according to the rule $(x^m)^n = x^{mn}$ from Sec. 6.5.

Example 5. $\sqrt{x^9} = \sqrt{x^8}\sqrt{x} = x^4\sqrt{x}$

Note: $x^8 x = x^8 x^1 = x^{8+1} = x^9$ according to the rule $x^m x^n = x^{m+n}$ from Sec. 6.5.

Directions: Factor any perfect squares out of each square root.

1) $\sqrt{9x} =$ 2) $\sqrt{5x^2} =$

3) $\sqrt{11x^2} =$ 4) $\sqrt{18x} =$

5) $\sqrt{6x^2} =$ 6) $\sqrt{36x} =$

7) $\sqrt{20x} =$ 8) $\sqrt{49x^2} =$

9) $\sqrt{x^4} =$ 10) $\sqrt{81x} =$

11) $\sqrt{8x^2} =$ 12) $\sqrt{x^3} =$

13) $\sqrt{x^{10}} =$ 14) $\sqrt{x^5} =$

If there is a square root in a denominator, it is customary to multiply both the numerator and denominator of the fraction by the (same) factor needed to remove the square root from the denominator. It's okay to have a square root in the numerator. For example, $\frac{1}{\sqrt{x}} = \frac{1}{\sqrt{x}}\frac{\sqrt{x}}{\sqrt{x}} = \frac{\sqrt{x}}{x}$ because $\sqrt{x}\sqrt{x} = x$.

Example 6. $\frac{1}{\sqrt{2}} = \frac{1}{\sqrt{2}}\frac{\sqrt{2}}{\sqrt{2}} = \frac{\sqrt{2}}{2}$

Example 7. $\frac{2}{\sqrt{x}} = \frac{2}{\sqrt{x}}\frac{\sqrt{x}}{\sqrt{x}} = \frac{2\sqrt{x}}{x}$

Example 8. $\frac{1}{\sqrt{3x}} = \frac{1}{\sqrt{3x}}\frac{\sqrt{3x}}{\sqrt{3x}} = \frac{\sqrt{3x}}{3x}$

Example 9. $\frac{\sqrt{5}}{\sqrt{x}} = \frac{\sqrt{5}}{\sqrt{x}}\frac{\sqrt{x}}{\sqrt{x}} = \frac{\sqrt{5x}}{x}$

Note: $\sqrt{5}\sqrt{x} = \sqrt{5x}$.

Directions: Rationalize each denominator.

15) $\frac{1}{\sqrt{3}} =$

16) $\frac{1}{\sqrt{2x}} =$

17) $\frac{x}{\sqrt{5}} =$

18) $\frac{3}{\sqrt{x}} =$

19) $\frac{\sqrt{2}}{\sqrt{x}} =$

20) $\frac{\sqrt{x}}{\sqrt{3}} =$

21) $\dfrac{6}{\sqrt{2x}} =$

22) $\dfrac{\sqrt{2}}{\sqrt{3}} =$

23) $\dfrac{1}{x\sqrt{x}} =$

24) $\dfrac{x}{\sqrt{x}} =$

25) $\dfrac{6}{\sqrt{2}} =$

26) $\dfrac{1}{2\sqrt{2}} =$

27) $\dfrac{x^2}{\sqrt{x}} =$

28) $\dfrac{6}{\sqrt{3}} =$

29) $\dfrac{10x}{\sqrt{5}} =$

30) $\dfrac{x}{\sqrt{7x}} =$

31) $\dfrac{14}{\sqrt{7x}} =$

7 SOLVING EQUATIONS

Consider the simple equation $3 = 3$. If we add 2 to both sides of this equation, we get $5 = 5$. If we now multiply both sides of the equation by 4, we get $20 = 20$. When we apply the same operation to both sides of an equation, the equation is still true.

However, if you only apply an operation to one side of an equation, it won't be true anymore. For example, starting with $3 = 3$, if you multiply only the left side by 2, the left side would be 6 while the right side would still be 3, and $6 \neq 3$.

If an equation contains a **variable** (like x), we can solve the equation by applying the same operation to both sides of the equation. For example, for $x + 4 = 12$, we can solve for x by subtracting 4 from both sides of the equation to get $x = 12 - 4 = 8$. You can check that this works by plugging $x = 8$ into the original equation: $8 + 4 = 12$.

Many equations can be solved by **isolating the unknown**. In order to isolate the unknown, do the same thing to both sides of the equation in such a way as to leave the variable all by itself on one side of the equation. The solution below shows an example of isolating the unknown.

$$5 + 2x = 23 - x \quad \text{(subtract 5 from both sides)}$$
$$2x = 18 - x \quad \text{(add } x \text{ to both sides)}$$
$$3x = 18 \quad \text{(divide both sides by 3)}$$
$$x = \frac{18}{3} = 6$$

Here is a summary of the logic behind the above solution:

- Our goal is to get the variable on just one side of the equation and the constants on the other side. We decided to bring the constants to the right and the variables to the left.
- First, we subtracted 5 from both sides of the equation so that there wouldn't be any constant terms on the left side.
- Next, we added x to both sides so that there wouldn't be any variables on the right.
- Once the variable term was isolated, we divided both sides by its coefficient (3).
- In each step, we basically did the **opposite** of what was being done. Since 5 was being added to $2x$, we subtracted 5 from both sides. Since, x was being subtracted from 18, we added x to both sides. Since x was being multiplied by 3, we divided both sides by 3. Note how each step does the opposite of what was being done.

Check the answer by plugging $x = 6$ into the original equation. On the left side, we get $5 + 2x = 5 + 2(6) = 5 + 12 = 17$. On the right side, we get $23 - x = 23 - 6 = 17$. Since both sides equal 17, the solution checks out.

7.1 One-Step Equations

The equations of this section are simple enough that they can be solved in a single step. Do the opposite of what is being done to the variable.

- If a constant is being added to a variable, subtract the constant from both sides. For example, subtract 2 from both sides of $x + 2 = 7$ to get $x = 7 - 2 = 5$.
- If a constant is being subtracted from a variable, add the constant to both sides. For example, add 5 to both sides of $x - 5 = 3$ to get $x = 3 + 5 = 8$.
- If a variable is being multiplied by a constant, divide both sides by the coefficient. For example, divide by 3 on both sides of $3x = 12$ to get $x = \frac{12}{3} = 4$.
- If a variable is being divided by a constant, multiply both sides by the denominator. For example, multiply by 4 on both sides of $\frac{x}{4} = 5$ to get $x = 4(5) = 20$.

Check your answer by plugging the answer into the original equation. For example, plug $x = 5$ into $x + 2 = 7$ to get $5 + 2 = 7$. Since $7 = 7$, the solution checks out.

Example 1. $x + 6 = 15$ Subtract 6 from both sides to get $x = 15 - 6 = 9$. Check: $9 + 6 = 15$.

Example 2. $x - 5 = 7$ Add 5 to both sides to get $x = 5 + 7 = 12$. Check: $12 - 5 = 7$.

Example 3. $2x = 8$ Divide both sides by 2 to get $x = \frac{8}{2} = 4$. Check: $2(4) = 8$.

Example 4. $\frac{x}{3} = 6$ Multiply both sides by 3 to get $x = 3(6) = 18$. Check: $\frac{18}{3} = 6$.

Directions: Solve for the variable as described above. Check your answer.

1) $x + 8 = 15$

2) $x - 7 = 4$

3) $7x = 63$

4) $\frac{x}{6} = 9$

5) $x - 5 = 5$

6) $9x = 81$

7) $\frac{x}{4} = 6$

8) $x + 9 = 12$

9) $4x = 52$

10) $x - 8 = 1$

11) $x + 5 = 11$

12) $\frac{x}{2} = 10$

13) $x - 6 = 10$

14) $6x = 42$

15) $\frac{x}{7} = 7$

16) $x + 3 = 9$

17) $x - 4 = 9$

18) $\frac{x}{3} = 9$

19) $10x = 100$

20) $x + 7 = 15$

21) $x - 9 = 13$

22) $\frac{x}{8} = 4$

23) $x + 6 = 10$

24) $5x = 60$

25) $\frac{x}{9} = 3$

26) $x - 3 = 12$

27) $8x = 56$

28) $x + 4 = 14$

29) $x - 10 = 20$

30) $\frac{x}{10} = 6$

7.2 Isolate the Unknown

Many equations can be solved by isolating the unknown as follows:
- Bring all of the constant terms to one side of the equation and all of the variable terms to the other side of the equation. If a term is being added to one side of the equation, you can move it to the other side by subtracting it from both sides. If a term is being subtracted, you can move it to the other side by adding it to both sides.
- Combine like terms (Sec. 6.2). For example, $9x - 3x = 6x$.
- Once you have a single variable term and a single constant term, divide both sides of the equation by the coefficient of the variable term (as we did in Sec. 7.1). For example, divide both sides of $4x = 20$ by 4 to get $x = \frac{20}{4} = 5$.

To check your answer, plug the value in for the unknown in the original equation. If both sides equal the same value, the answer checks out.

Example 1. $3x - 4 = 14$ We will put both constant terms on the right side.
- Add 4 to both sides: $3x - 4 + 4 = 14 + 4$
- Simplify. The 4 cancels on the left: $3x = 18$
- Divide by 3 on both sides: $\frac{3x}{3} = \frac{18}{3}$
- Simplify. The 3 cancels on the left: $x = 6$

Plug $x = 6$ into the original equation to check the answer: $3(6) - 4 = 18 - 4 = 14$.

Example 2. $2x + 25 = 7x$ We will put both variable terms on the right side.
- Subtract $2x$ from both sides: $2x + 25 - 2x = 7x - 2x$
- Combine like terms. The $2x$ cancels on the left: $25 = 5x$
- Divide by 5 on both sides: $\frac{25}{5} = \frac{5x}{5}$
- Simplify. The 5 cancels on the right: $5 = x$ (equivalent to $x = 5$)

Plug $x = 5$ into the original equation: $2x + 25 = 2(5) + 25 = 35$ agrees with $7(5) = 35$.

Example 3. $4x = 21 - 3x$ We will put both variable terms on the left side.
- Add $3x$ to both sides: $4x + 3x = 21 - 3x + 3x$
- Combine like terms. The $3x$ cancels on the right: $7x = 21$
- Divide by 7 on both sides: $\frac{7x}{7} = \frac{21}{7}$
- Simplify. The 7 cancels on the left: $x = 3$

Plug $x = 3$ into the original equation: $4(3) = 12$ agrees with $21 - 3(3) = 21 - 9 = 12$.

Directions: Solve for the variable by isolating it. Check your answer.

1) $2x + 3 = 17$

2) $3x + 24 = 7x$

3) $5x = 12 + 2x$

4) $29 = 4x + 5$

5) $8x - 9 = 31$

6) $15 - x = 2x$

7) $6x = 72 - 3x$

8) $27 = 5x - 8$

9) $7 + x = 2x$

10) $5 + 3x = 8x$

11) $-3 + 4x = 17$

12) $16 = -4 + 2x$

Example 4. $9 + 2x = 5x - 6$ Should we collect the variables on the left or right side?

- If we put the variables on the left, we will get $2x - 5x = -3x$, which is negative (like Sec. 7.3), but if we put the variables on the right, we will get $5x - 2x = 3x$, which is positive. It is simpler to put the variables on the right side for this problem.
- Subtract $2x$ from both sides: $9 + 2x - 2x = 5x - 6 - 2x$
- Combine like terms. The $2x$ cancels on the left: $9 = 3x - 6$
- Add 6 to both sides: $9 + 6 = 3x - 6 + 6$
- Simplify. The 6 cancels on the right: $15 = 3x$
- Divide by 3 on both sides: $\frac{15}{3} = \frac{3x}{3}$
- Simplify. The 3 cancels on the right: $5 = x$ (equivalent to $x = 5$)

Plug $x = 5$ into the original equation: $9 + 2(5) = 9 + 10 = 19$ agrees with $5(5) - 6 = 25 - 6 = 19$.

Directions: Solve for the variable by isolating it. Check your answer.

13) $5x - 2 = 3x + 10$

14) $24 - x = 3x + 8$

15) $2 + 5x = 9 + 4x$

16) $3x + 38 = 8x - 7$

17) $6x - 4 = 9x - 28$

18) $11 + 2x = 6x + 3$

19) $15 - x = 3 + x$

20) $5 - 3x = 17 - 9x$

21) $9 + 2x = 8x - 3x$

22) $7 + 6x = 36 - 5$

23) $64 - 4x = 8 + 4x$

24) $-2x - 3 = -7x + 7$

7.3 Variables with Negative Signs

Some problems have negative answers and some problems require dividing by a negative number. Recall that the answer to a division problem is negative if there is one minus sign, but the answer is positive if there are two minus signs. For example, compare $\frac{-8}{2} = \frac{8}{-2} = -4$ with $\frac{-8}{-4} = 2$.

Sometimes, you can avoid a minus sign. For example, consider the equation $5 - x = x + 1$. If you bring the variables to the left, you get a negative coefficient: $5 - 2x = 1$. However, if you bring the variables to the right, you get a positive coefficient: $5 = 2x + 1$. In this case, you can choose to avoid a negative coefficient.

Other times, it is more convenient to just deal with the minus sign. For example, consider the equation $1 - 2x = 9$. In this case, you can solve the problem in fewer steps if you leave the variable term on the left: $-2x = 8$ and then $x = \frac{8}{-2} = -4$. Even if you move the variable term to the right, you still have to deal with the minus sign: $1 = 9 + 2x$ then $-8 = 2x$ and $-\frac{8}{2} = x$.

Example 1. $5 - 3x = 14$ It is convenient to leave the variable term on the left side.
- Subtract 5 from both sides: $5 - 3x - 5 = 14 - 5$
- Simplify. The 5 cancels on the left: $-3x = 9$
- Divide by -3 on both sides: $\frac{-3x}{-3} = \frac{9}{-3}$
- Simplify. The -3 cancels on the left: $x = -3$

Plug $x = -3$ into the original equation to check the answer: $5 - 3(-3) = 5 + 9 = 14$.

Example 2. $4x - 16 = 8x$ It is convenient to put both variables on the right side.
- Subtract $4x$ from both sides: $4x - 16 - 4x = 8x - 4x$
- Simplify. The $4x$ cancels on the left: $-16 = 4x$
- Divide by 4 on both sides: $\frac{-16}{4} = \frac{4x}{4}$
- Simplify. The 4 cancels on the right: $-4 = x$ (equivalent to $x = -4$)

Plug $x = -4$ into the original equation to check the answer: $4(-4) - 16 = -16 - 16 = -32$ agrees with $8(-4) = -32$.

Directions: Solve for the variable by isolating it. Check your answer.

1) $7 - 2x = 13$

2) $7x = 4x - 12$

3) $9x = -35 + 4x$

4) $2 = 12 - 5x$

5) $28 + 6x = 4$

6) $-18 + 3x = -3x$

7) $5x + 14 = 3x$

8) $40 - 8x = -32$

9) $44 = 12 - 4x$

10) $-7x = 20 - 3x$

11) $2x = 9x - 56$

12) $-1 = 2x - 5$

Sometimes it is helpful to multiply or divide both sides of an equation by negative one. For example, this is the simplest way to solve for x in the equation $-x = -8$. Note that multiplying and dividing by -1 have the same result. For example, if we multiply both sides of $-x = -8$ by -1 we get $(-1)(-x) = (-1)(-8)$ which becomes $x = 8$, and if we divide both sides of $-x = -8$ by -1 we get $\frac{-x}{-1} = \frac{-8}{-1}$ which also becomes $x = 8$. Some instructors tell their students to multiply both sides by -1, while other instructors tell their students to divide both sides by -1, but it doesn't make any difference since the outcome will be the same either way.

When you multiply (or divide) both sides of an equation by -1, remember to apply this to every term in the equation. For example, $-2 = 5 - x$ would become $2 = -5 + x$. The effect is to change the sign of every term in the equation.

Example 3. $3 - x = -7$ Multiply every term of the equation by -1. This will change the sign of every term in the equation. The equation is now $-3 + x = 7$.
- Add 3 to both sides: $-3 + x + 3 = 7 + 3$
- Simplify. The 3 cancels on the left: $x = 10$

Plug $x = 10$ into the original equation to check the answer: $3 - 10 = -7$.

Directions: Solve for the variable by isolating it. Check your answer.

13) $4 = -x$

14) $1 - x = -2$

15) $-8 = -x - 4$

16) $-x = -9$

17) $2x + 3 = x$

18) $5 = 6 - x$

7.4 Variables with Fractions

Fractions are common in algebra. For example, sometimes the problem includes fractions, and sometimes the answer turns out to be a fraction. It may help to review Sec. 3.3 (making a common denominator in order to add or subtract fractions) and Sec. 3.6 (dividing fractions). When the answer is a fraction, it is customary to reduce the fraction (Sec. 3.1), if possible. For example, $\frac{12}{15}$ can be reduced by dividing by 3 in the numerator and denominator: $\frac{12 \div 3}{15 \div 3} = \frac{4}{5}$.

Example 1. $6x + 2 = 12$ It is convenient to leave the variable term on the left side.

- Subtract 2 from both sides: $6x + 2 - 2 = 12 - 2$
- Simplify. The 2 cancels on the left: $6x = 10$
- Divide by 6 on both sides: $\frac{6x}{6} = \frac{10}{6}$
- Simplify. The 6 cancels on the left: $x = \frac{10}{6}$
- Reduce the fraction. Divide by 2 in the numerator and denominator: $x = \frac{10 \div 2}{6 \div 2} = \frac{5}{3}$

Plug $x = \frac{5}{3}$ into the original equation to check the answer: $6\left(\frac{5}{3}\right) + 2 = \frac{30}{3} + 2 = 10 + 2 = 12$.

Example 2. $\frac{3}{4} = 3x$ It is convenient to leave the variable on the right side.

- Divide by 3 on both sides: $\frac{3}{4} \div 3 = \frac{3x}{3}$
- Simplify. The 3 cancels on the right: $\frac{1}{4} = x$ (since $\frac{3}{4} \div 3 = \frac{3}{4} \cdot \frac{1}{3} = \frac{3}{12} = \frac{3 \div 3}{12 \div 3} = \frac{1}{4}$)

Plug $x = \frac{1}{4}$ into the original equation to check the answer: $3\left(\frac{1}{4}\right) = \frac{3}{4}$.

Example 3. $2x + \frac{1}{3} = \frac{1}{2}$ It is convenient to leave the variable term on the left side.

- Subtract $\frac{1}{3}$ from both sides: $2x + \frac{1}{3} - \frac{1}{3} = \frac{1}{2} - \frac{1}{3}$
- Simplify. The $\frac{1}{3}$ cancels on the left: $2x = \frac{1}{6}$ (since $\frac{1}{2} - \frac{1}{3} = \frac{1 \cdot 3}{2 \cdot 3} - \frac{1 \cdot 2}{3 \cdot 2} = \frac{3}{6} - \frac{2}{6} = \frac{1}{6}$)
- Divide by 2 on both sides: $\frac{2x}{2} = \frac{1}{6} \div 2$
- Simplify. The 2 cancels on the left: $x = \frac{1}{12}$ (since $\frac{1}{6} \div 2 = \frac{1}{6} \cdot \frac{1}{2} = \frac{1}{12}$)

Plug $x = \frac{1}{12}$ into the original equation to check the answer: $2\left(\frac{1}{12}\right) + \frac{1}{3} = \frac{2}{12} + \frac{1 \cdot 4}{3 \cdot 4} = \frac{2}{12} + \frac{4}{12} = \frac{6}{12}$
$= \frac{6 \div 6}{12 \div 6} = \frac{1}{2}$.

Directions: Solve for the variable by isolating it. Check your answer.

1) $5x - 1 = 3$

2) $15 = 3 + 9x$

3) $3x = 2$

4) $\frac{8}{9} = 2x$

5) $x - \frac{7}{8} = \frac{5}{2}$

6) $\frac{5}{6} = 3x + \frac{1}{3}$

7) $10 + 2x = -4x$

8) $5 - 9x = -2$

9) $4x = -3$

10) $4 - 7x = 0$

11) $\frac{4}{5} + x = 2$

12) $2x - \frac{2}{3} = \frac{3}{4}$

7.5 Roots and Exponents of Variables

If you can isolate the square root of a variable (using the method from Sec. 7.2), after isolating the square root, square both sides of the equation. For example, for $\sqrt{x} = 3$, square both sides of the equation: $\left(\sqrt{x}\right)^2 = 3^2$. Recall from Sec. 6.12 that $\left(\sqrt{x}\right)^2 = \sqrt{x}\sqrt{x} = x$. Thus, the solution to $\sqrt{x} = 3$ is $x = 9$. Check the answer by plugging it into the original equation: $\sqrt{9} = 3$.

If you isolate a variable and it has an integer exponent (x^n), take the nth root of both sides of the equation. Recall from Sec. 1.8 that if the power is even (like x^2 or x^4), the answer to a root problem has two possible answers, indicated by a $\pm$ sign. For example, for $x^2 = 16$, square root both sides of the equation to get $\sqrt{x^2} = \sqrt{16}$, which then becomes $x = \pm 4$. Check the answers: $(-4)^2 = 16$ and $4^2 = 16$. If instead the power is odd (like x^3 or x^5), there is only one real answer. For example, for $x^3 = 27$, cube root both sides of the equation to get $\sqrt[3]{x^3} = 27$, which then becomes $x = 3$. Check the answer: $3^3 = 3 \cdot 3 \cdot 3 = 9 \cdot 3 = 27$.

Example 1. $3 + \sqrt{x} = 15$ Subtract 3 from both sides to get $\sqrt{x} = 12$.

- Square both sides: $\left(\sqrt{x}\right)^2 = 12^2$.
- The left side is $\left(\sqrt{x}\right)^2 = \sqrt{x}\sqrt{x} = x$.
- The right side is $12^2 = 12 \cdot 12 = 144$.
- The solution is $x = 144$.

Check the answer by plugging it into the original equation: $3 + \sqrt{144} = 3 + 12 = 15$.

Example 2. $x^2 - 6 = 30$ Add 6 to both sides to get $x^2 = 36$. Square root both sides: $\sqrt{x^2} = \sqrt{36}$.

- The left side is $\sqrt{x^2} = x$.
- The right side is $\sqrt{36} = \pm 6$.
- Since x^2 is an even power, the solution is $x = \pm 6$.

Check the answers by plugging them into the original equation: $(-6)^2 - 6 = 36 - 6 = 30$ and $6^2 - 6 = 36 - 6 = 30$.

Example 3. $2x^3 = 16$ Divide both sides by 2 to get $x^3 = 8$. Cube root both sides: $\sqrt[3]{x^3} = \sqrt[3]{8}$.

- The left side is $\left(\sqrt[3]{x}\right)^3 = x$.
- The right side is $\sqrt[3]{8} = 2$.
- The solution is $x = 2$.

Check the answer by plugging it into the original equation: $2(2)^3 = 2(8) = 16$.

Directions: Solve for the variable by isolating it. Check your answer.

1) $6 = \sqrt{x}$

2) $49 = x^2$

3) $x^3 = 64$

4) $\sqrt{x} = -8$

5) $\sqrt{x} - 2 = 5$

6) $4x^4 = 1024$

7) $2x^2 = 50$

8) $12 = 3\sqrt{x}$

9) $2 - \sqrt{x} = 0$

10) $x^2 - 9 = 16$

11) $25 - x^3 = 150$

12) $8\sqrt{x} = 72$

7.6 Variables in a Denominator

If the variable is in the denominator of a fraction, first isolate the variable term (following the method of Sec. 7.2). Simplify the constant side of the equation down to a single term. Then take the reciprocal of both sides (recall Sec. 3.5). For example, for $\frac{1}{x} = \frac{3}{4}$, when we take the reciprocal of both sides, we get $x = \frac{4}{3}$.

Example 1. $\frac{1}{x} - \frac{1}{4} = \frac{1}{12}$ It is convenient to leave the variable term on the left side.

- Add $\frac{1}{4}$ to both sides: $\frac{1}{x} - \frac{1}{4} + \frac{1}{4} = \frac{1}{12} + \frac{1}{4}$
- Simplify. The $\frac{1}{4}$ cancels on the left: $\frac{1}{x} = \frac{1}{3}$ (since $\frac{1}{12} + \frac{1}{4} = \frac{1}{12} + \frac{1 \cdot 3}{4 \cdot 3} = \frac{1}{12} + \frac{3}{12} = \frac{4}{12} = \frac{1}{3}$)
- Take the reciprocal of both sides: $x = \frac{3}{1} = 3$

Plug $x = 3$ into the original equation to check the answer: $\frac{1}{3} - \frac{1}{4} = \frac{1 \cdot 4}{3 \cdot 4} - \frac{1 \cdot 3}{4 \cdot 3} = \frac{4}{12} - \frac{3}{12} = \frac{1}{12}$.

Directions: Solve for the variable by isolating it. Check your answer.

1) $\frac{1}{x} + \frac{1}{9} = \frac{1}{6}$

2) $\frac{11}{12} = \frac{5}{4} - \frac{1}{x}$

3) $\frac{5}{3} - \frac{1}{x} = \frac{13}{15}$

4) $\frac{1}{x} + \frac{5}{6} = \frac{7}{8}$

7.7 Cross Multiplying

When two fractions are equal and at least one fraction involves a variable, the problem can be solved by cross multiplying. This means to multiply along the diagonals as shown below.

$$\frac{w}{x} = \frac{y}{z} \quad \rightarrow \quad \frac{w}{x} \diagup\hspace{-1em}\diagdown \frac{y}{z} \quad \rightarrow \quad wz = xy$$

The numerator of the first fraction multiplies the denominator of the second fraction, and the denominator of the first fraction multiplies the numerator of the second fraction. For example, for $\frac{3}{x} = \frac{6}{8}$, cross multiplying gives $3(8) = 6x$, which becomes $24 = 6x$. We can solve this to get $4 = x$. You can check that this works because $\frac{6}{8} = \frac{6\div2}{8\div2} = \frac{3}{4}$.

Example 1. $\frac{x}{3} = \frac{4}{5}$ Cross multiply to get $5x = 3(4)$, which becomes $5x = 12$. Divide both sides by 5 to get $x = \frac{12}{5}$.

Plug $x = \frac{12}{5}$ into the original equation to check the answer: $\frac{12}{5} \div 3 = \frac{12}{5} \cdot \frac{1}{3} = \frac{12}{15} = \frac{12\div3}{15\div3} = \frac{4}{5}$.

Example 2. $\frac{5}{x} = \frac{2}{3}$ Cross multiply to get $5(3) = 2x$, which becomes $15 = 2x$. Divide both sides by 2 to get $\frac{15}{2} = x$ (equivalent to $x = \frac{15}{2}$).

Plug $x = \frac{15}{2}$ into the original equation to check the answer: $5 \div \frac{15}{2} = \frac{5}{1} \cdot \frac{2}{15} = \frac{10}{15} = \frac{10\div5}{15\div5} = \frac{2}{3}$.

Directions: Solve for the variable by isolating it. Check your answer.

1) $\dfrac{x}{4} = \dfrac{12}{16}$

2) $\dfrac{2}{x} = \dfrac{8}{12}$

3) $\dfrac{3}{4} = \dfrac{x}{2}$

4) $\dfrac{8}{3} = \dfrac{5}{x}$

5) $\frac{3}{8} = \frac{x}{6}$

6) $\frac{5}{x} = \frac{15}{2}$

7) $\frac{x}{7} = \frac{2}{21}$

8) $\frac{5}{4} = \frac{10}{x}$

9) $\frac{9}{2} = \frac{3}{2x}$

10) $\frac{3x}{2} = \frac{15}{4}$

11) $\frac{5}{8x} = \frac{15}{4}$

12) $\frac{1}{7} = \frac{4x}{35}$

13) $\frac{3}{4} = \frac{6x}{8}$

14) $\frac{4}{5x} = \frac{8}{25}$

7.8 Special Solutions

Some equations don't have a solution. For example, $x - x = 2$ has no solution. For any value of x, the left side equals zero, and zero will never be equal to two.

For some equations, every real number solves the equation. For example, $3x = 3x$ is true for all real numbers. This time when x cancels out, we get $3 = 3$ which is always true.

Example 1. $x + 1 = x$ Subtract x from both sides: $1 = 0$. That can't be true. This equation has no solution.

Example 2. $2x + 3x = 5x$ Combine like terms: $5x = 5x$. Divide by x on both sides: $5 = 5$. This is always true. The solution to this equation is all real numbers.

Directions: Determine whether the answer is no solution or all real numbers.

1) $x + x = 2x$

2) $x + 2 = x + 3$

3) $5 - x^2 = 3 - x^2$

4) $x^2 - x^2 = 0$

5) $5x = 1 + 5x$

6) $x + 1 - x = 0$

7) $1 + x - 1 = x$

8) $\dfrac{x}{2} = \dfrac{2x}{4}$

8 RATIO PROBLEMS

A <u>ratio</u> expresses a fixed relationship in the form $x{:}y$ (with a colon between two numbers). For example, if 125 girls and 100 boys attend a school, the ratio of girls to boys is 125:100. This ratio can be <u>reduced</u> to 5:4 the same way that the fraction $\frac{125}{100}$ can be reduced to $\frac{5}{4}$: By dividing the numerator and denominator each by the greatest common factor: $\frac{125}{100} = \frac{125 \div 25}{100 \div 25} = \frac{5}{4}$. There are three basic types of ratios relating the parts and the whole. For example, consider the English alphabet, which has 26 letters in total, including 21 consonants and 5 vowels. In this example, the <u>parts</u> are 21 consonants and 5 vowels, and the whole is 26 letters. The parts make up the <u>whole</u>.

- Part to part. For example, the ratio of consonants to vowels, 21:5, is part to part.
- Part to whole. For example, the ratio of vowels to letters, 5:26, is part to whole.
- Whole to part. For example, the ratio of letters to vowels, 26:5, is whole to part.

Ratios can also be expressed as fractions or percents. For example, on an exam suppose that a student solved 7 problems correctly for every 3 problems that the student solved incorrectly. In this example, the ratio of correct answers to incorrect answers is 7:3. As a fraction, we could express this as $\frac{7}{3}$. However, it is more common to express a fraction or a percent using a part and a whole. Add 7 and 3 together to determine that the whole is 10. The ratio of questions answered correctly in part-to-whole form is 7:10. In fractional form, $\frac{7}{10}$ of the questions were answered correctly. As a percent, 70% of the questions were answered correctly.

A unit ratio is an equivalent ratio where one of the numbers (typically, the second number) is equal to one. For example, for the ratio 3:2, a unit ratio of 1.5:1 can be formed (by dividing each number by 2). Unit ratios make it easy to compare different ratios. For example, which ratio is bigger, 9:5 or 5:2? Divide each ratio by the second number to make unit ratios. In this case, we get 1.8:1 and 2.5:1 (since $\frac{9}{5} = \frac{9 \cdot 2}{5 \cdot 2} = \frac{18}{10} = 1.8$ and $\frac{5}{2} = \frac{5 \cdot 5}{2 \cdot 5} = \frac{25}{10} = 2.5$, where the middle dot represents multiplication whereas the lower dot is a decimal point). Since $1.8 < 2.5$, the ratio 9:5 is less than the ratio 5:2. This means that 5:2 is bigger than 9:5.

A ratio can actually consist of more than two numbers. For example, a cube has 8 corners, 12 edges, and 6 faces. The ratio of its corners to edges to faces is 8:12:6. We could separate this ratio to say that the ratio of corners to edges is 8:12 and the ratio of edges to faces is 12:6, or even that the ratio of corners to faces is 8:6.

8.1 Reducing Ratios

To reduce a ratio, divide each number by the greatest common factor. For example, the 8 and 12 of 8:12 are both multiples of 4, allowing us to write:

$$8{:}12 = \frac{8}{12} = \frac{8 \div 4}{12 \div 4} = \frac{2}{3} = 2{:}3$$

If you divide by a common factor that isn't the GCF, you'll need to divide by another common factor in order to completely reduce the ratio. For example, consider $\frac{24 \div 6}{36 \div 6} = \frac{4}{6}$ which can be reduced further by $\frac{4 \div 2}{6 \div 2} = \frac{2}{3}$. We could achieve this in one step with $\frac{24 \div 12}{36 \div 12} = \frac{2}{3}$.

Example 1. $15{:}20 = \frac{15}{20} = \frac{15 \div 5}{20 \div 5} = \frac{3}{4} = 3{:}4$ $\qquad$ **Example 2.** $9{:}6 = \frac{9}{6} = \frac{9 \div 3}{6 \div 3} = \frac{3}{2} = 3{:}2$

1) $4{:}16 =$

2) $12{:}18 =$

3) $14{:}6 =$

4) $4{:}18 =$

5) $12{:}16 =$

6) $8{:}24 =$

7) $8{:}2 =$

8) $14{:}38 =$

9) $16{:}24 =$

10) $22{:}34 =$

11) $40{:}15 =$

12) $26{:}4 =$

13) $18{:}24 =$

14) $21{:}48 =$

15) $44{:}8 =$

16) $30{:}35 =$

17) $34{:}12 =$

18) $42{:}52 =$

8.2 The Part and the Whole

It's important to pay attention to the words that describe a ratio. You need to know whether either of the given numbers is the whole or if both numbers are parts. For example, consider the word RATIO, which has 2 consonants, 3 vowels, and 5 letters. We can make different types of ratios out of this word:

- The ratio of consonants to vowels is 2:3.
- The ratio of consonants to letters is 2:5.
- The ratio of letters to vowels is 5:3.

In this example, the 3 consonants are one part, the 2 vowels are another part, and the 5 letters are the whole. Note that the two parts add up to the whole. In this case, $2 + 3 = 5$.

Example 1. There are 12 cars and 8 trucks in a parking lot.

- The total number of vehicles is $12 + 8 = 20$. We add since the given numbers are parts.
- The ratio of cars to trucks is 3:2 (since $\frac{12}{8} = \frac{12 \div 4}{8 \div 4} = \frac{3}{2}$).
- The ratio of cars to vehicles is 3:5 (since $\frac{12}{20} = \frac{12 \div 4}{20 \div 4} = \frac{3}{5}$).

Example 2. The ratio of red apples to total apples in a barrel is 4:7. The other apples are green.

- To make a ratio using green apples, subtract the red apples from the total apples in the given ratio: $7 - 4 = 3$. We subtract since the given ratio included the whole.
- The ratio of red apples to green apples is 4:3.

1) For the word PERCENTAGE, express the following ratios in reduced form:

(a) What is the ratio of vowels to consonants?

(b) What is the ratio of consonants to letters?

(c) What is the ratio of letters to vowels?

2) The ratio of children to adults at a gathering is 3:4. Express the ratio of people to adults as a reduced ratio.

3) For the number 314159265, express the following ratios in reduced form:

(a) What is the ratio of the total number of digits to the odd digits?

(b) What is the ratio of odd digits to even digits?

(c) What is the ratio of even digits to the total number of digits?

4) The ratio of pencils to writing utensils in a container is 3:4. The other writing utensils are pens. Express the ratio of pens to pencils as a reduced ratio.

5) There are 32 dogs and 40 cats at a kennel. Express the following ratios in reduced form:

(a) What is the ratio of cats to dogs?

(b) What is the ratio of dogs to pets?

(c) What is the ratio of pets to cats?

6) The ratio of softcover books to hardcover books at a library is 3:7. Express the ratio of hardcover books to the total number of books as a reduced ratio.

7) There are 48 people in a room. 18 are male. Express the following ratios in reduced form:

(a) What is the ratio of females to people?

(b) What is the ratio of females to males?

8.3 Ratios and Fractions

To find a fraction from a ratio (or vice-versa), pay attention to the wording. For example, if the ratio of girls to boys is 5:4, the fraction of the students that are boys is $\frac{4}{9}$ (since the whole is $4 + 5 = 9$ students).

Example 1. The ratio of two-story houses to one-story houses in a community is 2:5. What fraction of these houses have two stories?

- 2:5 is part to part. The whole is $2 + 5 = 7$.
- The fraction of houses with two stories is $\frac{2}{7}$.

Example 2. Three-fourths of the students did their homework. What is the ratio of students who did their homework to the students who didn't do their homework?

- $\frac{3}{4}$ is part to whole. The other part (those who didn't do their homework) is $4 - 3 = 1$.
- 3 students did their homework for every 1 who didn't. The desired ratio is 3:1.

Example 3. The ratio of snacks to healthy snacks in a vending machine is 7:3. What fraction of these snacks aren't considered healthy?

- 7:3 is whole to part. The other part (those that aren't healthy) is $7 - 3 = 4$.
- The fraction of snacks that aren't healthy is $\frac{4}{7}$.

1) The ratio of voters to nonvoters in a district was 5:3. What fraction of these people voted?

2) Two-fifths of the customers used a coupon. What is the ratio of customers who used a coupon to those who didn't use a coupon?

3) The ratio of students who didn't earn an A on the test to those who earned an A on the test is 17:3. What fraction of the students earned an A on the test?

4) Three-sevenths of the tickets sold at a movie theater were children's tickets. What is the ratio of adult tickets to children's tickets?

8.4 Ratios and Percents

To find a percent from a ratio (or vice-versa), first find the fraction like we did in Sec. 8.3. It may help to review Sec.'s 5.3 and 5.4 (converting between fractions and percents).

Example 1. The ratio of customers to employees in a store is 18:7. What percent of these people are customers?
- 18:7 is part to part. The whole is $18 + 7 = 25$.
- The percent who are customers is $\frac{18}{25} = \frac{18 \cdot 4}{25 \cdot 4} = \frac{72}{100} = 72\%$.

Example 2. Of the lunches served at a restaurant, 20% are vegetarian. What is the ratio of non-vegetarian lunches to vegetarian lunches?
- Convert the percent to a fraction: $20\% = \frac{20}{100} = \frac{20 \div 20}{100 \div 20} = \frac{1}{5}$.
- $\frac{1}{5}$ is part to whole. The other part (non-vegetarian) is $5 - 1 = 4$.
- 4 lunches are non-vegetarian for every 1 that is vegetarian. The desired ratio is 4:1.

Example 3. The ratio of cars to vehicles in a parking lot is 37:50. What percent of these vehicles aren't cars?
- 37:50 is part to whole. The other part (those that aren't cars) is $50 - 37 = 13$.
- The percent that aren't cars is $\frac{13}{50} = \frac{13 \cdot 2}{50 \cdot 2} = \frac{26}{100} = 26\%$.

1) The ratio of moms to dads at a parent group is 3:2. What percent of the parents are moms?

2) Of the flowers sold at a floral shop, 75% were roses. What is the ratio of roses sold to other types of flowers sold?

3) The ratio of girls to students in a class is 13:20. What percent of these students are boys?

4) Of the cars that drove by in the past minute, 18% had their headlights on. What is the ratio of the cars with their headlights off to the total number of cars that drove by?

8.5 Unit Ratios

A <u>unit ratio</u> (also known as unitary ratio) is a ratio where one of the numbers (typically, the second number) is equal to one. For example, 4:1 and 1.5:1 are unit ratios. In this section, we will make the second number equal to one. It may help to review Sec. 4.5 (converting fractions to decimals).

Example 1. $7:2 = \frac{7}{2} = \frac{7 \cdot 5}{2 \cdot 5} = \frac{35}{10} = 3.5:1$ **Example 2.** $3:4 = \frac{3}{4} = \frac{3 \cdot 25}{4 \cdot 25} = \frac{75}{100} = 0.75:1$

Directions: Convert each ratio to a unit fraction where the second number is equal to one.

1) $8:5 =$

2) $7:20 =$

3) $9:4 =$

4) $1:2 =$

5) $17:50 =$

6) $32:25 =$

7) $15:8 =$

8) $283:100 =$

9) $53:10 =$

10) $19:200 =$

11) $7:40 =$

12) $11:4 =$

13) $1:5 =$

14) $521:250 =$

15) $999:500 =$

16) $207:125 =$

8.6 Linear Proportions

When two quantities are linearly proportional, an increase in one of the quantities results in an increase in the other quantity by the same factor (and similarly for a decrease in one of the quantities). For example, if 3 jars can hold 100 balls, then 12 jars can hold 400 balls. Note that $3 \times 4 = 12$ and $100 \times 4 = 400$ (that is, both quantities increased by a factor of 4).

One way to solve a problem involving a linear proportion is to follow these steps:
- Use x to represent a quantity that it may help to solve for.
- Express two ratios in the same way. For example, for a ratio of girls to boys, each ratio should be a fraction with girls in the numerator and boys in the denominator.
- Set the two fractions equal to one another. For example, if the ratio of girls to boys is 6:5 and the total number of boys is 40, the equation would be $\frac{6}{5} = \frac{x}{40}$. Note that 6 and x both correspond to girls, while 5 and 40 both correspond to boys.
- Cross multiply. It may help to review Sec. 7.7. For the previous example, $6(40) = 5x$.
- Isolate the unknown (Sec. 7.2). Here, $240 = 5x$ becomes $\frac{240}{5} = x$ and $48 = x$.
- Does the answer make sense? For example, for the ratio of 6 girls to 5 boys, there are more girls than boys. This agrees with 48 girls and 40 boys.

Example 1. The ratio of apples to oranges in a barrel is 3:5. There are 18 apples in the barrel. How many oranges are in the barrel?
- Let x represent the number of oranges in the barrel.
- Two ratios of apples to oranges are 3:5 and 18:x. Set these fractions equal: $\frac{3}{5} = \frac{18}{x}$.
- Cross multiply: $3x = 5(18)$. Simplify: $3x = 90$. Isolate the unknown: $x = \frac{90}{3} = 30$.
- The ratio 3:5 has fewer apples than oranges. 18 apples is less than 30 oranges.

Example 2. The ratio of working lightbulbs to lightbulbs in a building is 9:10. There are 80 lightbulbs in the building. How many lightbulbs need to be replaced?
- Let x represent the number of working lightbulbs. Note: This will **not** be the answer.
- Two ratios of working lightbulbs to total lightbulbs are 9:10 and x:80. Set these fractions equal: $\frac{9}{10} = \frac{x}{80}$.

- Cross multiply: $9(80) = 10x$. Simplify: $720 = 10x$. Isolate the unknown: $\frac{720}{10} = 72 = x$.
- 72 out of 80 are working. The final answer is that $80 - 72 = 8$ need to be replaced.
- The ratio 9:10 has more working lightbulbs than not. 72 is greater than 8.

1) The ratio of girls to boys at a dance is 7:6. There are 54 boys at the dance. How many girls are at the dance?

2) A store sold computers. The ratio of laptops sold to desktops sold was 5:2. The store sold 750 laptops. How many desktops did the store sell?

3) The ratio of defective televisions to the total number of televisions is 3:20. The total number of televisions is 600. How many of the televisions are defective?

4) The ratio of seniors who graduated from a high school to the total number of seniors was 7:10. The total number of seniors is 300. How many seniors didn't graduate?

5) The ratio of parents to children at a gathering is 4:9. There are 28 parents at the gathering. How many children are at the gathering?

6) The ratio of apples to red apples in a barrel is 8:3. There are 56 apples in the barrel. How many of these are red?

7) A school uses 12 buses to transport 600 students. At this rate, how many buses would be needed to transport 150 of the students on a field trip?

8) The ratio of ripe tomatoes to the total number of tomatoes in a field is 7:9. There are 720 tomatoes in the field. How many of the tomatoes aren't ripe?

9 RATE PROBLEMS

A **rate** is made by dividing two quantities that have different units. For example, $\frac{200 \text{ miles}}{5 \text{ hours}}$ is a rate formed by dividing 200 miles by 5 hours. Since $200 \div 5 = 40$, this rate is equivalent to 40 mph (where mph stands for miles per hour). As another example, if you read 5 pages in 4 minutes, this rate is $\frac{5 \text{ pages}}{4 \text{ min.}}$. Since $5 \div 4 = 1.25$, this rate is equivalent to 1.25 pages per min. Rates that don't turn out to be whole numbers are typically expressed as decimals instead of fractions. For example, if a woman has a job that pays \$140 for 8 hours of work, the woman would be more likely to tell you that she earns \$17.50 per hour than to tell you that she earns \$$\frac{140}{8}$ per hour (or even the reduced fraction \$$\frac{35}{2}$ per hour).

One rate that is very common in math and physics problems is speed. The **speed** of an object provides a measure of how fast it is moving. For example, a car that travels 30 m/s is moving faster than a car that travels 25 m/s. If a car travels with a constant speed of 30 m/s, every second the car travels 30 meters.

For an object that travels with **constant speed**, the speed of the object is related to the distance traveled and elapsed time by the following equation:

$$r = \frac{d}{t}$$

- r represents the speed (which is a rate). Note: Some books use v instead (for velocity).
- d represents the distance traveled. Note: Some books use x instead.
- t represents the elapsed time. This is the time that the object has been traveling.

If you need to solve for distance or time, the above formula can be rewritten as:

$$d = rt \quad , \quad t = \frac{d}{r}$$

Note that d is never on the bottom of this equation.

When using the rate equation, be sure that the distance, rate, and time have compatible units. For example, if the rate is in mph (miles per hour), the distance needs to be in miles and the time needs to be in hours.

The process of expressing a quantity in different units is called a unit **conversion**. For example, since 1 yard $=$ 3 feet, we can convert 4 yards into 12 feet. Similarly, since 1 hour $=$ 60 minutes, we can convert 5 hours into 300 minutes. We will explore unit conversions in Sec.'s 9.3-9.4.

9.1 Rates and Decimals

Convert each rate to a decimal. It may help to review Sec. 4.5 (converting fractions to decimals.)

Example 1. 28 miles in 5 hours $= \frac{28 \text{ miles}}{5 \text{ hours}} = \frac{28 \times 2 \text{ miles}}{5 \times 2 \text{ hours}} = \frac{56 \text{ miles}}{10 \text{ hours}} = 5.6$ mph

Example 2. 9 cents for 4 buttons $= \frac{9 \text{ cents}}{4 \text{ buttons}} = \frac{9 \times 25 \text{ cents}}{4 \times 25 \text{ buttons}} = \frac{225 \text{ cents}}{100 \text{ buttons}} = 2.25$ cents per button

Note: The denominator changes from plural to singular after converting to decimal. It costs 9 cents for 4 buttons (plural), which is an average cost of 2.25 cents for one button (singular).

Directions: Convert each rate to a decimal.

1) 81 meters in 20 seconds =

2) 7 pages in 2 minutes =

3) 1300 calories in 5 servings =

4) 4 inches in 25 days =

5) 18 kilometers in 4 hours =

6) 11 grams in 50 milliliters =

7) 87 dollars for 10 tickets =

8) 260 miles for 8 gallons =

9.2 Constant Speed

If an object travels with constant speed (r), its speed can be found by dividing the distance traveled (d) by the elapsed time (t):

$$r = \frac{d}{t}$$

If you need to solve for distance, multiply the speed by the time. (You can get the equation below by multiplying both sides of the previous equation by the time.)

$$d = rt$$

If you need to solve for time, divide the distance by the speed. (You can get the equation below by dividing both sides of the previous equation by the speed.)

$$t = \frac{d}{r}$$

In all three forms of the equation, note that distance is never in the denominator. Remember this and it may help you avoid making a mistake. Although distance and time may be measured in different units, the units of distance, time, and speed must be consistent. For example, if the speed is in mph (miles per hour), the distance needs to be in miles and the time needs to be in hours. If an object travels with a constant speed of 30 mph, it travels 30 miles in 1 hour, 60 miles in 2 hours, 90 miles in 3 hours, 120 miles in 4 hours, etc. These numbers agree with the equation for constant speed. For example, $\frac{60 \text{ miles}}{2 \text{ hours}} = 30$ mph and $\frac{90 \text{ miles}}{3 \text{ hours}} = 30$ mph. The units can help you determine which quantities are given in a problem. The speed will have a slash (/) or the word per in it, like m/s or mph. The distance is a length like m, mi., ft., yd., in., etc. The time is a duration like s, hr., min., day, yr., etc.

Example 1. An airplane travels 400 mi. in 5 hr. What is the speed of the airplane?
- $d = 400$ mi. is the distance traveled (in miles).
- $t = 5$ hr. is the elapsed time (in hours).
- We're solving for the speed (r). To find speed, divide the distance by the time.

$$r = \frac{d}{t} = \frac{400 \text{ mi.}}{5 \text{ hr.}} = 80 \text{ mph}$$

Example 2. A car travels 30 m/s for 8 s. How far does the car travel?
- $r = 30$ m/s is the speed (in meters per second).
- $t = 8$ s is the elapsed time (in seconds).

- We're solving for the distance (d). To find distance, multiply the speed by the time.

$$d = rt = (30 \text{ m/s})(8 \text{ s}) = 240 \text{ m}$$

1) A girl rides a bicycle with a constant speed of 25 m/s for 7 s. How far does she travel?

2) What is the speed of a train that travels 600 ft. in 5 s?

3) If a boy travels at a steady rate of 3 m/s, how long will it take the boy to travel 150 m?

4) How fast does an ambulance need to drive in order to travel 270 miles in 3 hours?

5) A bug travels 7 in./min. for 9 min. How far does the bug travel?

6) How much time does it take to travel 300 km at a constant speed of 50 km/hr?

7) A wheel rolls with constant speed, traveling 72 cm in 9 s. What is the speed of the wheel?

9.3 Unit Conversions

If a distance is measured to be 6 feet and the relationship 1 ft. = 12 in. is used to express the distance as 72 inches, the units are said to have been converted from feet to inches. One way to convert units is to multiply by a fraction that equals one. Multiplying by one doesn't change the value. If the numerator equals the denominator, the fraction equals one. For example, the fraction $\frac{12 \text{ in.}}{1 \text{ ft.}}$ is equal to one because 12 in. is the same as 1 ft. If we multiply 6 ft. by this fraction, the ft. will cancel and the distance will be converted from feet to inches:

$$6 \text{ ft.} = 6 \text{ ft.} \times \frac{12 \text{ in.}}{1 \text{ ft.}} = 72 \text{ in.}$$

To perform a unit conversion, follow these steps:

- First you need the proper conversion factor, like 1 km = 1000 m or 1 hr. = 60 s.
- Make a fraction that is equal to one using the two numbers from each conversion factor. Choose the numerator and denominator to cancel the desired units. For example, in $6 \text{ ft.} \times \frac{12 \text{ in.}}{1 \text{ ft.}}$ the ft. cancel, but $6 \text{ ft.} \times \frac{1 \text{ ft.}}{12 \text{ in.}}$ is wrong since there is no cancellation.
- Some problems may require multiple conversion factors. See Example 3.
- Multiply the original value by the fraction(s). To carry out the arithmetic, multiply by any numerators and divide by any denominators.

Example 1. Convert 35,000 m to kilometers, given that 1 km = 1000 m.

Put 1000 m in the denominator to cancel the meters.

$$35{,}000 \text{ m} = 35{,}000 \text{ m} \times \frac{1 \text{ km}}{1000 \text{ m}} = 35 \text{ km}$$

We divided by 1000 because it is in the denominator.

Example 2. Convert 3 min. to seconds, given that 1 min. = 60 s.

Put 1 min. in the denominator to cancel the minutes.

$$3 \text{ min.} = 3 \text{ min.} \times \frac{60 \text{ s}}{1 \text{ min.}} = 180 \text{ s}$$

We multiplied by 60 because it is in the numerator.

Example 3. Convert 5 yds. to inches, given that 1 yd. = 3 ft. and 1 ft. = 12 in.

Put 1 yd. in the denominator to cancel the yards. Then put 1 ft. in the denominator to cancel the feet.

$$5 \text{ yds.} = 5 \text{ yds.} \times \frac{3 \text{ ft.}}{1 \text{ yd.}} \times \frac{12 \text{ in.}}{1 \text{ ft.}} = 180 \text{ in.}$$

We multiplied by 3 and 12 because these are both in the numerator.

1) Convert 8 ft. to inches, given that 1 ft. = 12 in.

2) Convert 21 ft. to yards, given that 1 yd. = 3 ft.

3) Convert 240 min. to hours, given that 1 hr. = 60 min.

4) Convert 15 kg to grams, given that 1 kg = 1000 g.

5) Convert 12 gal. to quarts, given that 1 gal. = 4 qt.

6) Convert 96 hr. to days, given that 1 day = 24 hr.

7) Convert 42 cm to mm, given that 1 cm = 10 mm.

8) Convert 4 hr. to seconds, given that 1 hr. = 60 min. and 1 min. = 60 s.

9) Convert 252 in. to yards, given that 1 yd. = 3 ft. and 1 ft. = 12 in.

9.4 Converting the Units of Rates

When the given units have a ratio, follow these steps to convert the units:

- Write the given units as a fraction with a horizontal line. For example, write $\frac{\text{mi.}}{\text{hr.}}$ instead of mph and write $\frac{\text{m}}{\text{s}}$ instead of m/s.

- To cancel units in the numerator, the denominator of the conversion fraction needs to have the same units as the old numerator. For example, in $5\frac{\text{m}}{\text{s}} \times \frac{1\text{ km}}{1000\text{ m}}$ the m cancels.

- To cancel units in the denominator, the numerator of the conversion fraction needs to have the same units as the old denominator. For example, in $5\frac{\text{m}}{\text{s}} \times \frac{60\text{ s}}{1\text{ min.}}$ the s cancels.

- Multiply numbers in numerators and divide by numbers in denominators.

Example 1. Convert 2 m/s to km/hr., given that 1 km = 1000 m and 1 hr. = 3600 s.

$$2\frac{\text{m}}{\text{s}} = 2\frac{\text{m}}{\text{s}} \times \frac{1\text{ km}}{1000\text{ m}} \times \frac{3600\text{ s}}{1\text{ hr.}} = \frac{2 \times 3600}{1000}\frac{\text{km}}{\text{hr.}} = 7.2\text{ km/hr.}$$

Example 2. Convert 10 ft./min. to in./s, given that 1 ft. = 12 in. and 1 min. = 60 s.

$$10\frac{\text{ft.}}{\text{min.}} = 10\frac{\text{ft.}}{\text{min.}} \times \frac{12\text{ in.}}{1\text{ ft.}} \times \frac{1\text{ min.}}{60\text{ s.}} = \frac{10 \times 12}{60}\frac{\text{in.}}{\text{s.}} = 2\text{ in./s.}$$

1) Convert 108 km/hr. to m/s, given that 1 km = 1000 m and 1 hr. = 3600 s.

2) Convert 8 qt./s to gal/min., given that 1 gal. = 4 qt. and 1 min. = 60 s.

9.5 Rate Formulas

Many rate formulas have a structure that is similar to the constant speed equation. Given a formula, sometimes you need to isolate a different symbol. For example, suppose that you are given the following formula, but wish to solve for q.

$$I = \frac{q}{t}$$

In that case, you could multiply both sides of the equation by t (using techniques from Chapter 7). You would get $It = q$, which is equivalent to:

$$q = It$$

Example 1. Given $P = \frac{W}{t}$, isolate W. Multiply both sides by t to get $Pt = W$, which is equivalent to $W = Pt$.

Example 2. Given $E = vB$, isolate v. Divide both sides by B to get $\frac{E}{B} = v$, which is equivalent to $v = \frac{E}{B}$.

Example 3. Given $a = \frac{v}{t}$, isolate t. First multiply both sides by t to get $at = v$. Now divide both sides by a to get $t = \frac{v}{a}$.

1) Given $V = IR$, isolate I.

2) Given $C = \frac{Q}{V}$, isolate Q.

3) Given $M = -\frac{q}{p}$, isolate q.

4) Given $d = \frac{m}{V}$, isolate V.

5) Given $d = abc$, isolate b.

6) Given $f = \frac{1}{T}$, isolate T.

7) Given $\frac{xy}{z} = 1$, isolate y.

8) Given $h = \frac{1}{2}gt^2$, isolate g.

9.6 Rates

Some simple rate formulas are much like the equation for constant speed, except that some other quantity is used besides distance or time. Such formulas can be deduced by examining the units. For example, suppose that a student reads 3 pages per minute. The "per" indicates that 3 pages per minute is the rate. The units indicate that the pages read divided by the time equals the rate.

$$\text{rate} = \frac{\text{pages}}{\text{minutes}}$$

How many pages would the student read in 12 minutes? To solve for pages, multiply both sides of the equation by time.

$$\text{pages} = (\text{rate})(\text{minutes}) = (3)(12) = 36$$

Example 1. A car is rated 50 mpg (miles per gallon). How far can the car travel on 12 gallons of gas? Note that 50 mpg is the rate.

$$\text{rate} = \frac{\text{miles}}{\text{gallons}}$$

To solve for the distance (in miles), multiply both sides of the equation by gallons.

$$\text{miles} = (\text{rate})(\text{gallons}) = (50)(12) = 600$$

Example 2. Water enters a tank at a rate of 4 liters per minute. The tank can hold 20 liters of water. How much time will it take for the tank to fill?

$$\text{rate} = \frac{\text{liters}}{\text{minutes}}$$

Multiply both sides of the equation by the time (in minutes).

$$(\text{rate})(\text{minutes}) = \text{liters}$$

Now divide both sides of the equation by the rate.

$$\text{minutes} = \frac{\text{liters}}{\text{rate}} = \frac{20}{4} = 5$$

1) A person can build 6 machines in one day. How many machines can the person build in one week if the person works every day, including the weekend?

2) If a student reads 7 pages per minute, how long does it take for the student to read 56 pages?

3) A car can travel 600 miles on 15 gallons of gas. How many miles per gallon does the car get?

4) A woman earns $30 per hour. How much does she earn in 5 days, working 8 hours per day?

5) A man wishes to save $750 in 10 months. How much money should he save each month?

6) A tank holding 20 gallons of water develops a leak at a rate of 0.25 gallons per minute. How much time will it take for the tank to become empty?

10 INEQUALITIES

The symbol $>$ is a **greater than sign**, while the symbol $<$ is a **less than sign**. Below are a few examples of inequalities that use these signs.

$$4 < 5 \quad , \quad 3 > 0 \quad , \quad -3 < -1$$

These read as "4 is less than 5," "3 is greater than 0," and "negative 3 is less than negative 1." Note that if you swap the values on the two sides of an inequality, the direction of the inequal sign changes. For example, compare $4 < 5$ with $5 > 4$; these two inequalities are effectively equivalent.

Less than and greater than symbols can be used to make comparisons. For example, suppose that a question asks, "Which is greater, 3.16 or 3.61?" Since 3.61 is greater, the answer can be stated as $3.16 < 3.61$. (It could also be stated as $3.61 > 3.16$.) The vertex (the point where the two line segments are joined) of the less than or greater than sign points to the smaller number.

When comparing two negative numbers, the negative number that is closer to 0 is larger. For example, let's compare -2 and -3. Which of these is closer to 0? The answer is -2. Another way to think of this is that -2 is less negative than -3, which makes -2 larger (it is closer to being positive). The answer can be expressed as $-2 > -3$ (negative 2 is greater than negative 3). Alternatively, it can be expressed as $-3 < -2$ (negative 3 is less than negative 2).

If one of the numbers is positive and the other isn't positive, the positive number is always larger. For example, $1 > -99$.

However, if **absolute values** are used, then you must ignore any minus signs. Two vertical lines around a number indicate an absolute value. For example, $|-4|$ means "the absolute value of negative 4" and is equal to 4. That is, it means to ignore the minus sign: $|-4| = 4$. Taking the absolute value of a positive number (or even zero) has no effect: $|5| = 5$. To compare numbers where the negative numbers are in absolute values, simply ignore the signs. For example, $|-7|$ is greater than 6, such that $|-7| > 6$.

One way to compare fractions is to find a **common denominator**. For example, to compare $\frac{2}{3}$ with $\frac{3}{5}$, we can make a common denominator of 15. We get $\frac{2}{3} = \frac{2 \cdot 5}{3 \cdot 5} = \frac{10}{15}$ and $\frac{3}{5} = \frac{3 \cdot 3}{5 \cdot 3} = \frac{9}{15}$. Since $\frac{10}{15} > \frac{9}{15}$, it follows that $\frac{2}{3} > \frac{3}{5}$.

Inequalities can also involve variables. For example, $3x > 12$ is an inequality. We can isolate the unknown of an inequality using the method of Sec. 7.2, with an important exception that involves negative coefficients (which we will explore in Sec. 10.7).

10.1 Negative Numbers and Absolute Values

To compare two negative numbers, whichever negative number is closer to zero is larger. Put another way, the less negative number is larger. For example, -4 is greater than -5 since -4 is closer to zero (it is less negative). We can express this as $-4 > -5$ (negative 4 is greater than negative 5).

If only one number is negative, the other number is automatically larger. For example, -10 is less than 0 (since zero isn't negative). We can express this as $-10 < 0$ (negative 10 is less than zero).

The straight lines surrounding $|-7|$ make an absolute value. To find the absolute value of a number, ignore any minus sign inside of the absolute value lines (if there is one). For example, $|-7|$ means the same thing as 7. Taking the absolute value of a positive number or zero has no effect. For example, $|9|$ is equal to 9.

Apply any absolute values before comparing numbers. For example, to compare $|-8|$ with 3, first note that $|-8| = 8$, and then compare 8 with 3. Since $8 > 3$, it follows that $|-8| > 3$.

Example 1. $-5 > -7$ (since -5 is not as negative as -7, it follows that -5 is greater than -7)

Example 2. $-8 < -4$ (since -8 is more negative than -4, it follows that -8 is less than -4)

Example 3. $-6 < 3$ (since -6 is negative, it must be less than any positive number)

Example 4. $|-9| > 2$ (since $|-9| = 9$ and $9 > 2$)

Example 5. $|-1| < 10$ (since $|-1| = 1$ and $1 < 10$)

Directions: Decide whether to insert $<$ or $>$ between each pair of values.

1) -23	-19		2) -11	-17	
3) 0	-1		4) 89	90	
5) -101	-99		6) -34	26	
7) $|-2|$	-3		8) 13	$|-12|$	
9) 29	$|-31|$		10) $|-8|$	$|-9|$	
11) $|-33|$	$|-25|$		12) $|-4|$	0	

10.2 Comparing Decimals and Percents

To compare two decimal values:

- First look at the integer parts. If one number has a larger integer part than the other, then that number is bigger. For example, $6.3 > 5.8$ because $6 > 5$.
- If the integer parts are equal, compare the tenths place to see which number is larger. For example, $3.7 < 3.9$ because the integer parts are the same and because $.7 < .9$.
- If the integer parts are equal and the tenths place is the same, compare the hundredths place. For example, $8.15 > 8.13$ since both numbers start with 8.1 and since $.05 > .03$.
- If necessary, go onto the thousandths place, and so on, until a digit is different.

If one number is an integer, you can add .0 to it. For example, 3 and 3.0 are the same.

If you need another digit after the decimal point, you can add trailing zeroes. For example, 1.570 and 1.5700 are the same as 1.57.

If one number is a percent, divide by 100% to convert the percent into a decimal (Sec. 5.2). If any numbers are negative, the ideas from Sec. 10.1 also apply. For example, $-2.7 > -2.9$ because -2.7 is closer to zero (it is not as negative as -2.9).

Example 1. $4.3 > 3.7$ (since $4 > 3$)   **Example 2.** $1.948 < 2.043$ since $1 < 2$

Example 3. $8.5 < 8.6$ (since the 8 is the same, $.5 < .6$ determines which is greater)

Example 4. $3.15 > 3.14$ (since 3.1 is the same, $.05 > .04$ determines which is greater)

Example 5. $5 < 5.01$ (since 5.00 is the same as 5 and since $5.00 < 5.01$)

Example 6. $0.38 > 35\%$ (since $35\% = \frac{35\%}{100\%} = 0.35$ and since $0.38 > 0.35$)

Directions: Decide whether to insert $<$ or $>$ between each pair of values.

1) 0.7	0.4		2) 9.5	9.6	
3) 2.736	1.949		4) 3.14	3.1	
5) 7.234	7.237		6) 0.29	25%	
7) 0.5	30%		8) 1.25	210%	
9) −0.4	−0.6		10) −0.13	−0.11	

10.3 Comparing Fractions

In order to compare two fractions that have different denominators, multiply the numerator and denominator of each fraction by the factor needed to make a common denominator (like we did in Sec. 3.3 to add or subtract fractions). Once the fractions have a common denominator, you can simply compare their numerators. For example, to compare $\frac{3}{7}$ to $\frac{4}{11}$, make a common denominator of 77. Multiply the numerator and denominator of $\frac{3}{7}$ each by 11 to get $\frac{3}{7} = \frac{3 \cdot 11}{7 \cdot 11} = \frac{33}{77}$ and multiply the numerator and denominator of $\frac{4}{11}$ each by 7 to get $\frac{4}{11} = \frac{4 \cdot 7}{11 \cdot 7} = \frac{28}{77}$. Since $\frac{33}{77} > \frac{28}{77}$, it follows that $\frac{3}{7} > \frac{4}{11}$.

Example 1. $\frac{3}{2} > \frac{4}{3}$ since $\frac{3}{2} = \frac{3 \cdot 3}{2 \cdot 3} = \frac{9}{6}, \frac{4}{3} = \frac{4 \cdot 2}{3 \cdot 2} = \frac{8}{6}$, and $\frac{9}{6} > \frac{8}{6}$
Alternate solution. Cross multiply: $3(3) > 2(4)$ since $9 > 8$
Example 2. $\frac{5}{6} < \frac{7}{8}$ since $\frac{5}{6} = \frac{5 \cdot 4}{6 \cdot 4} = \frac{20}{24}, \frac{7}{8} = \frac{7 \cdot 3}{8 \cdot 3} = \frac{21}{24}$, and $\frac{20}{24} < \frac{21}{24}$
Alternate solution. Cross multiply: $5(8) < 6(7)$ since $40 < 42$

Directions: Decide whether to insert $<$ or $>$ between each pair of values.

1) $\frac{1}{2}$ $\frac{3}{7}$

2) $\frac{5}{3}$ $\frac{7}{5}$

3) $\frac{1}{4}$ $\frac{1}{5}$

4) $\frac{5}{6}$ $\frac{8}{9}$

5) $-\frac{4}{3}$ $-\frac{5}{4}$

10.4 Comparing Rational Numbers

Rational numbers include integers like 22 and fractions like $\frac{5}{3}$. Since decimals and percents can be converted into fractions, they are also rational. To compare rational numbers, convert all of the numbers to the same form (such as decimal) and then compare the numbers. It may help to review Sec.'s 4.5 and 4.6 (converting fractions to decimals).

Example 1. $\frac{5}{4} < 1.3$ since $\frac{5}{4} = \frac{5 \cdot 25}{4 \cdot 25} = \frac{125}{100} = 1.25$ and since $1.25 < 1.3$

Example 2. $\frac{11}{5} > 2$ since $\frac{11}{5} = \frac{11 \cdot 2}{5 \cdot 2} = \frac{22}{10} = 2.2$ and since $2.2 > 2$

Example 3. $70\% > \frac{2}{3}$ since $70\% = \frac{70\%}{100\%} = 0.7, \frac{2}{3} = \frac{2 \cdot 3}{3 \cdot 3} = \frac{6}{9} = 0.\overline{6}$, and $0.7 > 0.\overline{6}$

Note: $0.\overline{6}$ is a repeating decimal, $0.6666666 \cdots$ with the 6 repeating forever (Sec. 4.6)

Directions: Decide whether to insert $<$ or $>$ between each pair of values.

1) 3.7 $\qquad$ $\frac{7}{2}$

2) $\frac{9}{20}$ $\qquad$ 0.42

3) 35% $\qquad$ $\frac{4}{11}$

4) 3 $\qquad$ $\frac{11}{4}$

5) 0.567 $\qquad$ $\frac{3}{5}$

6) -1.6 $\qquad$ $-\frac{7}{4}$

10.5 Comparing Rational and Irrational Numbers

The square root of a number that isn't a perfect square (like $\sqrt{3}$) is irrational. (In contrast, $\sqrt{4}$ is rational because $\sqrt{4} = 2$. The difference is that 4 is a perfect square, since $2^2 = 4$.) To compare a rational number to a square root, follow these steps:

- Square the rational number. For example, $1.4^2 = (1.4)(1.4) = 1.96$ (Sec. 4.10).
- Write the rational number as the square root of itself squared. For example, $1.4 = \sqrt{1.96}$.
- Compare the values inside of the square roots. For example, since $\sqrt{1.96} < \sqrt{2}$, it follows that $1.4 < \sqrt{2}$.

Example 1. $2.5 > \sqrt{6}$ since $2.5 = \sqrt{(2.5)(2.5)} = \sqrt{6.25}$ and since $\sqrt{6.25} > \sqrt{6}$

Example 2. $3 < \sqrt{10}$ since $3 = \sqrt{(3)(3)} = \sqrt{9}$ and since $\sqrt{9} < \sqrt{10}$

Directions: Decide whether to insert $<$ or $>$ between each pair of values.

1) 2 $\qquad$ $\sqrt{3}$

2) 2.2 $\qquad$ $\sqrt{5}$

3) $\sqrt{17}$ $\qquad$ 4

4) 5.4 $\qquad$ $\sqrt{30}$

5) $\sqrt{79}$ $\qquad$ 9

6) $\sqrt{0.3}$ $\qquad$ $\frac{1}{4}$

10.6 Inequalities with Variables

For many inequalities, the variable can be isolated using the method from Sec. 7.2. (We will explore an important exception in Sec. 10.7.) For example, the variable in $4x + 6 < 18$ can be isolated by subtracting 6 from both sides to get $4x < 12$ and dividing by 3 on both sides to get $x < 3$. We can check that this makes sense using a calculator. For example, plug in 2.9 for x to see that $4(2.9) + 6 = 11.6 + 6 = 17.6$ is indeed a little less than 18, whereas if you plug in 3.1 for x you get $4(3.1) + 6 = 12.4 + 6 = 18.4$ is greater than 18. These agree with $x < 3$.

Example 1. $5x - 8 > 12$ We will put both constant terms on the right side.

- Add 8 to both sides: $5x - 8 + 8 > 12 + 8$
- Simplify. The 8 cancels on the left: $5x > 20$
- Divide by 5 on both sides: $\frac{5x}{5} > \frac{20}{5}$. Simplify. The 5 cancels on the left: $x > 4$

Plug 4.1 in for x to see that $5(4.1) - 8 = 20.5 - 8 = 12.5$ is greater than 12. Plug 3.9 in for x to see that $5(3.9) - 8 = 19.5 - 8 = 11.5$ is less than 12. These agree with $x > 4$.

Directions: Solve for the variable by isolating it.

1) $2x + 7 < 15$

2) $3x + 6 < 9x$

3) $3 > 4x - 9$

4) $4x > 56 - 3x$

5) $4x > 7 + 3x$

6) $6x - 5 < 25$

10.7 Inequalities with Variables and Minus Signs

When dividing (or multiplying) both sides of an inequality by a negative number, the direction of the inequality reverses. For example, if you divide by -2 on both sides of $-2x > -8$, you get $x < -4$. Note how this changed the $>$ sign into a $<$ sign. Why does the sign change? As an example, consider the inequality $5 < 7$. If you multiply both sides of this inequality by minus one, you get -5 on the left and -7 on the right. Since -5 is greater than (less negative than) -7, in order for the inequality to be true, we must reverse the original sign: $-5 > -7$.

This rule **only** applies when multiplying or dividing both sides of an inequality by a **negative** number. It doesn't apply when multiplying or dividing by positive numbers. It doesn't apply when adding or subtracting numbers.

Example 1. $-x < -3$ Multiply (or divide) both sides of the inequality by -1. This will change the $<$ sign into a $>$ sign. The answer is $x > 3$.
Alternate solution: Add x to both sides to get $0 < -3 + x$ and add 3 to both sides to get $3 < x$.
Example 2. $-3x > 15$ Divide both sides by -3. This will change the $>$ sign into a $<$ sign. The answer is $x < -5$.
Alternate solution: Add $3x$ to both sides to get $0 > 15 + 3x$, subtract 15 from both sides to get $-15 > 3x$, and divide by 3 on both sides to get $-5 > x$, which is equivalent to $x < -5$.
Example 3. $3 - 2x > -9$ We will put both constant terms on the right side.

- Subtract 3 from both sides: $3 - 2x - 3 > -9 - 3$
- Simplify. The 3 cancels on the left: $-2x > -12$
- Divide by -2 on both sides. This changes the $>$ sign into a $<$ sign: $\frac{-2x}{-2} < \frac{-12}{-2}$.
- Simplify. The -2 cancels on the left: $x < 6$

Alternate solution: Add $2x$ to both sides to get $3 > -9 + 2x$, add 9 to both sides to get $12 > 2x$, and divide by 2 on both sides to get $6 > x$, which is equivalent to $x < 6$.

Directions: Solve for the variable by isolating it. It may help to review Sec.'s 7.6 and 7.7.

1) $8 > -x$

2) $-42 < -7x$

3) $1 - x < -1$

4) $15 > -9 - 3x$

5) $-4x + 8 < 28$

6) $-5 < -4x - 9$

7) $-3x > 10 - 2x$

8) $5x + 10 < -20$

9) $-18 + 3x > -6x$

10) $-36 - 4x > -8x$

11) $-x - 3 > -9 - 3x$

12) $-(-x) < -7$

13) $-\frac{x}{2} < \frac{5}{3}$

14) $-\frac{4}{x} > -\frac{3}{2}$

15) $\frac{1}{2} - \frac{x}{3} > \frac{1}{6}$

16) $\frac{1}{3} - \frac{1}{x} < \frac{1}{12}$

ANSWER KEY

Chapter 1 Exponents

1.1 Squares

1) $2^2 = 2 \times 2 = 4$

2) $5^2 = 5 \times 5 = 25$

3) $8^2 = 8 \times 8 = 64$

4) $1^2 = 1 \times 1 = 1$

5) $6^2 = 6 \times 6 = 36$

6) $3^2 = 3 \times 3 = 9$

7) $9^2 = 9 \times 9 = 81$

8) $0^2 = 0 \times 0 = 0$

9) $4^2 = 4 \times 4 = 16$

10) $10^2 = 10 \times 10 = 100$

11) $15^2 = 15 \times 15 = 225$

12) $25^2 = 25 \times 25 = 625$

13) $11^2 = 11 \times 11 = 121$

14) $14^2 = 14 \times 14 = 196$

15) $30^2 = 30 \times 30 = 900$

16) $18^2 = 18 \times 18 = 324$

17) $16^2 = 16 \times 16 = 256$

18) $12^2 = 12 \times 12 = 144$

19) $20^2 = 20 \times 20 = 400$

20) $19^2 = 19 \times 19 = 361$

1.2 Cubes

1) $3^3 = 3 \times 3 \times 3 = 9 \times 3 = 27$

2) $1^3 = 1 \times 1 \times 1 = 1 \times 1 = 1$

3) $7^3 = 7 \times 7 \times 7 = 49 \times 7 = 343$

4) $5^3 = 5 \times 5 \times 5 = 25 \times 5 = 125$

5) $10^3 = 10 \times 10 \times 10 = 100 \times 10 = 1000$

6) $2^3 = 2 \times 2 \times 2 = 4 \times 2 = 8$

7) $12^3 = 12 \times 12 \times 12 = 144 \times 12 = 1728$

8) $0^3 = 0 \times 0 \times 0 = 0 \times 0 = 0$

9) $15^3 = 15 \times 15 \times 15 = 225 \times 15 = 3375$

10) $9^3 = 9 \times 9 \times 9 = 81 \times 9 = 729$

11) $13^3 = 13 \times 13 \times 13 = 169 \times 13 = 2197$

12) $4^3 = 4 \times 4 \times 4 = 16 \times 4 = 64$

13) $20^3 = 20 \times 20 \times 20 = 400 \times 20 = 8000$

14) $8^3 = 8 \times 8 \times 8 = 64 \times 8 = 512$

1.3 Powers

1) $2^7 = 2 \times 2 \times 2 \times 2 \times 2 \times 2 \times 2 = 4 \times 4 \times 4 \times 2 = 16 \times 8 = 128$

2) $4^5 = 4 \times 4 \times 4 \times 4 \times 4 = 16 \times 16 \times 4 = 256 \times 4 = 1024$

3) $8^4 = 8 \times 8 \times 8 \times 8 = 64 \times 64 = 4096$

4) $6^5 = 6 \times 6 \times 6 \times 6 \times 6 = 36 \times 36 \times 6 = 1296 \times 6 = 7776$

5) $11^0 = 1$

6) $7^4 = 7 \times 7 \times 7 \times 7 = 49 \times 49 = 2401$

7) $2^9 = 2 \times 2 \times 2 \times 2 \times 2 \times 2 \times 2 \times 2 \times 2 = 4 \times 4 \times 4 \times 4 \times 2 = 16 \times 16 \times 2 = 512$

8) $9^4 = 9 \times 9 \times 9 \times 9 = 81 \times 81 = 6561$

9) $20^1 = 20$

10) $13^3 = 13 \times 13 \times 13 = 169 \times 13 = 2197$

11) $5^5 = 5 \times 5 \times 5 \times 5 \times 5 = 25 \times 25 \times 5 = 625 \times 5 = 3125$

12) $10^6 = 10 \times 10 \times 10 \times 10 \times 10 \times 10 = 100 \times 100 \times 100 = 10,000 \times 100 = 1,000,000$

13) $3^7 = 3 \times 3 \times 3 \times 3 \times 3 \times 3 \times 3 = 9 \times 9 \times 9 \times 3 = 81 \times 27 = 2187$

14) $100^2 = 100 \times 100 = 10,000$

1.4 Negative Numbers

1) $(-6)^3 = -6 \times (-6) \times (-6) = 36 \times (-6) = -216$

2) $(-5)^4 = -5 \times (-5) \times (-5) \times (-5) = 25 \times 25 = 625$

3) $(-2)^6 = -2 \times (-2) \times (-2) \times (-2) \times (-2) \times (-2) = 4 \times 4 \times 4 = 16 \times 4 = 64$

4) $(-7)^3 = -7 \times (-7) \times (-7) = 49 \times (-7) = -343$

5) $(-4)^6 = -4 \times (-4) \times (-4) \times (-4) \times (-4) \times (-4) = 16 \times 16 \times 16 = 256 \times 16 = 4096$

6) $(-10)^5 = -10 \times (-10) \times (-10) \times (-10) \times (-10) = 100 \times 100 \times (-10) = -100,000$

7) $(-8)^4 = -8 \times (-8) \times (-8) \times (-8) = 64 \times 64 = 4096$

8) $(-11)^3 = -11 \times (-11) \times (-11) = 121 \times (-11) = -1331$

9) $(-9)^4 = -9 \times (-9) \times (-9) \times (-9) = 81 \times 81 = 6561$

10) $(-12)^2 = -12 \times (-12) = 144$

11) $(-2)^7 = -2 \times (-2) \times (-2) \times (-2) \times (-2) \times (-2) \times (-2) = 4 \times 4 \times 4 \times (-2) = 64 \times (-2) = -128$

12) $(-3)^6 = -3 \times (-3) \times (-3) \times (-3) \times (-3) \times (-3) = 9 \times 9 \times 9 = 81 \times 9 = 729$

13) $(-20)^3 = -20 \times (-20) \times (-20) = 400 \times (-20) = -8000$

14) $(-5)^5 = -5 \times (-5) \times (-5) \times (-5) \times (-5) = 25 \times 25 \times (-5) = 625 \times (-5) = -3125$

1.5 Negative Powers

1) $6^{-1} = \frac{1}{6}$

2) $7^{-2} = \frac{1}{7^2} = \frac{1}{7 \times 7} = \frac{1}{49}$

3) $4^{-3} = \frac{1}{4^3} = \frac{1}{4 \times 4 \times 4} = \frac{1}{16 \times 4} = \frac{1}{64}$

4) $2^{-5} = \frac{1}{2^5} = \frac{1}{2 \times 2 \times 2 \times 2 \times 2} = \frac{1}{4 \times 4 \times 2} = \frac{1}{16 \times 2} = \frac{1}{32}$

5) $8^{-2} = \frac{1}{8^2} = \frac{1}{8 \times 8} = \frac{1}{64}$

6) $3^{-4} = \frac{1}{3^4} = \frac{1}{3 \times 3 \times 3 \times 3} = \frac{1}{9 \times 9} = \frac{1}{81}$

7) $10^{-3} = \frac{1}{10^3} = \frac{1}{10 \times 10 \times 10} = \frac{1}{100 \times 10} = \frac{1}{1000}$

8) $12^{-1} = \frac{1}{12}$

9) $5^{-4} = \frac{1}{5^4} = \frac{1}{5 \times 5 \times 5 \times 5} = \frac{1}{25 \times 25} = \frac{1}{625}$

10) $(-9)^{-2} = \frac{1}{(-9)^2} = \frac{1}{(-9) \times (-9)} = \frac{1}{81}$

11) $2^{-7} = \frac{1}{2^7} = \frac{1}{2 \times 2 \times 2 \times 2 \times 2 \times 2 \times 2} = \frac{1}{4 \times 4 \times 4 \times 2} = \frac{1}{16 \times 8} = \frac{1}{128}$

12) $(-3)^{-5} = \frac{1}{(-3)^5} = \frac{1}{(-3) \times (-3) \times (-3) \times (-3) \times (-3)} = \frac{1}{9 \times 9 \times (-3)} = \frac{1}{81 \times (-2)} = \frac{1}{-243} = -\frac{1}{243}$

13) $1^{-1} = \frac{1}{1^1} = \frac{1}{1} = 1$

14) $(-20)^{-1} = \frac{1}{(-20)^1} = \frac{1}{-20} = -\frac{1}{20}$

15) $8^{-3} = \frac{1}{8^3} = \frac{1}{8 \times 8 \times 8} = \frac{1}{64 \times 8} = \frac{1}{512}$

1.6 Square Roots

1) $\sqrt{16} = \pm 4$ since $(\pm 4)^2 = 4 \times 4 = 16$

2) $\sqrt{4} = \pm 2$ since $(\pm 2)^2 = 2 \times 2 = 4$

3) $\sqrt{225} = \pm 15$ since $(\pm 15)^2 = 15 \times 15 = 225$

4) $\sqrt{25} = \pm 5$ since $(\pm 5)^2 = 5 \times 5 = 25$

5) $\sqrt{100} = \pm 10$ since $(\pm 10)^2 = 10 \times 10 = 100$

6) $\sqrt{64} = \pm 8$ since $(\pm 8)^2 = 8 \times 8 = 64$

7) $\sqrt{625} = \pm 25$ since $(\pm 25)^2 = 25 \times 25 = 625$

8) $\sqrt{81} = \pm 9$ since $(\pm 9)^2 = 9 \times 9 = 81$

9) $\sqrt{289} = \pm 17$ since $(\pm 17)^2 = 17 \times 17 = 289$

10) $\sqrt{36} = \pm 6$ since $(\pm 6)^2 = 6 \times 6 = 36$

11) $\sqrt{144} = \pm 12$ since $(\pm 12)^2 = 12 \times 12 = 144$

12) $\sqrt{1} = \pm 1$ since $(\pm 1)^2 = 1 \times 1 = 1$

13) $\sqrt{196} = \pm 14$ since $(\pm 14)^2 = 14 \times 14 = 196$

14) $\sqrt{49} = \pm 7$ since $(\pm 7)^2 = 7 \times 7 = 49$

1.7 Cube Roots

1) $\sqrt[3]{27} = 3$ because $3^3 = 3 \times 3 \times 3 = 9 \times 3 = 27$

2) $\sqrt[3]{125} = 5$ because $5^3 = 5 \times 5 \times 5 = 25 \times 5 = 125$

3) $\sqrt[3]{64} = 4$ because $4^3 = 4 \times 4 \times 4 = 16 \times 4 = 64$

4) $\sqrt[3]{-216} = -6$ because $(-6)^3 = -6 \times (-6) \times (-6) = 36 \times (-6) = -216$

5) $\sqrt[3]{729} = 9$ because $9^3 = 9 \times 9 \times 9 = 81 \times 9 = 729$

6) $\sqrt[3]{-343} = -7$ because $(-7)^3 = -7 \times (-7) \times (-7) = 49 \times (-7) = -343$

7) $\sqrt[3]{1000} = 10$ because $10^3 = 10 \times 10 \times 10 = 100 \times 10 = 1000$

8) $\sqrt[3]{512} = 8$ because $8^3 = 8 \times 8 \times 8 = 64 \times 8 = 512$

1.8 Roots

1) $\sqrt[4]{256} = \pm 4$ because $(\pm 4)^4 = 4 \times 4 \times 4 \times 4 = 16 \times 16 = 256$

2) $\sqrt[3]{512} = 8$ because $8^3 = 8 \times 8 \times 8 = 64 \times 8 = 512$

3) $\sqrt[4]{625} = \pm 5$ because $(\pm 5)^4 = 5 \times 5 \times 5 \times 5 = 25 \times 25 = 625$

4) $\sqrt[4]{16} = \pm 2$ because $(\pm 2)^4 = 2 \times 2 \times 2 \times 2 = 4 \times 4 = 16$

5) $\sqrt[5]{-243} = -3$ because $(-3)^5 = -3 \times (-3) \times (-3) \times (-3) \times (-3) = 81 \times (-3) = -243$

6) $\sqrt[5]{100,000} = 10$ because $10^5 = 10 \times 10 \times 10 \times 10 \times 10 = 100 \times 1000 = 100,000$

7) $\sqrt[3]{-216} = -6$ because $(-6)^3 = -6 \times (-6) \times (-6) = 36 \times (-6) = -216$

8) $\sqrt[4]{4096} = \pm 8$ because $(\pm 8)^4 = 8 \times 8 \times 8 \times 8 = 64 \times 64 = 4096$

9) $\sqrt[4]{10,000} = \pm 10$ because $(\pm 10)^4 = 10 \times 10 \times 10 \times 10 = 100 \times 100 = 10,000$

10) $\sqrt[6]{64} = \pm 2$ because $(\pm 2)^6 = 2 \times 2 \times 2 \times 2 \times 2 \times 2 = 4 \times 4 \times 4 = 16 \times 4 = 64$

11) $\sqrt[6]{729} = \pm 3$ because $(\pm 3)^6 = 3 \times 3 \times 3 \times 3 \times 3 \times 3 = 9 \times 9 \times 9 = 81 \times 9 = 729$

12) $\sqrt[5]{-1024} = -4$ because $(-4)^5 = -4 \times (-4) \times (-4) \times (-4) \times (-4) = 256 \times (-4) = -1024$

13) $\sqrt[3]{343} = 7$ because $7^3 = 7 \times 7 \times 7 = 49 \times 7 = 343$

14) $\sqrt[9]{-512} = -2$ because $(-2)^9 = -2 \times (-2) \times (-2) \times (-2) \times (-2) \times (-2) \times (-2) \times (-2) \times (-2)$
$= 4 \times 4 \times 4 \times 4 \times (-2) = 16 \times 16 \times (-2) = 256 \times (-2) = -512$

15) $\sqrt[4]{1296} = \pm 6$ because $(\pm 6)^4 = 6 \times 6 \times 6 \times 6 = 36 \times 36 = 1296$

16) $\sqrt[3]{1,000,000} = 100$ because $100^3 = 100 \times 100 \times 100 = 10,000 \times 100 = 1,000,000$

17) $\sqrt[5]{-3125} = -5$ because $(-5)^5 = -5 \times (-5) \times (-5) \times (-5) \times (-5) = 625 \times (-5) = -3125$

18) $\sqrt[3]{-8000} = -20$ because $(-20)^3 = -20 \times (-20) \times (-20) = 400 \times (-20) = -8000$

1.9 Prime Factorization

1) $2^2 \times 3$ because $2^2 = 2 \times 2 = 4$ and $4 \times 3 = 12$

2) $3^2 \times 5$ because $3^2 = 3 \times 3 = 9$ and $9 \times 5 = 45$

3) $3^2 \times 11$ because $3^2 = 3 \times 3 = 9$ and $9 \times 11 = 99$

4) $2 \times 5 \times 7$ because $2 \times 5 \times 7 = 10 \times 7 = 70$

5) $5^2 \times 7$ because $5^2 = 5 \times 5 = 25$ and $25 \times 7 = 175$

6) $2^2 \times 7^2$ because $2^2 = 2 \times 2 = 4, 7^2 = 7 \times 7 = 49,$ and $4 \times 49 = 196$

7) 2×5^3 because $5^3 = 5 \times 5 \times 5 = 125$ and $2 \times 125 = 250$

8) $2^3 \times 3^2$ because $2^3 = 2 \times 2 \times 2 = 8, 3^2 = 3 \times 3 = 9,$ and $8 \times 9 = 72$

9) $2^2 \times 5 \times 11$ because $2^2 = 2 \times 2 = 4$ and $4 \times 5 \times 11 = 20 \times 11 = 220$

1.10 Factor Out Perfect Squares

1) $\sqrt{12} = \sqrt{4} \times \sqrt{3} = 2 \times \sqrt{3} = 2\sqrt{3}$

2) $\sqrt{63} = \sqrt{9} \times \sqrt{7} = 3 \times \sqrt{7} = 3\sqrt{7}$

3) $\sqrt{80} = \sqrt{16} \times \sqrt{5} = 4 \times \sqrt{5} = 4\sqrt{5}$

4) $\sqrt{108} = \sqrt{36} \times \sqrt{3} = 6 \times \sqrt{3} = 6\sqrt{3}$

5) $\sqrt{27} = \sqrt{9} \times \sqrt{3} = 3 \times \sqrt{3} = 3\sqrt{3}$

6) $\sqrt{128} = \sqrt{64} \times \sqrt{2} = 8 \times \sqrt{2} = 8\sqrt{2}$

7) $\sqrt{175} = \sqrt{25} \times \sqrt{7} = 5 \times \sqrt{7} = 5\sqrt{7}$

8) $\sqrt{48} = \sqrt{16} \times \sqrt{3} = 4 \times \sqrt{3} = 4\sqrt{3}$

9) $\sqrt{243} = \sqrt{81} \times \sqrt{3} = 9 \times \sqrt{3} = 9\sqrt{3}$

10) $\sqrt{44} = \sqrt{4} \times \sqrt{11} = 2 \times \sqrt{11} = 2\sqrt{11}$

11) $\sqrt{500} = \sqrt{100} \times \sqrt{5} = 10 \times \sqrt{5} = 10\sqrt{5}$

12) $\sqrt{98} = \sqrt{49} \times \sqrt{2} = 7 \times \sqrt{2} = 7\sqrt{2}$

13) $\sqrt{288} = \sqrt{144} \times \sqrt{2} = 12 \times \sqrt{2} = 12\sqrt{2}$

14) $\sqrt{320} = \sqrt{64} \times \sqrt{5} = 8 \times \sqrt{5} = 8\sqrt{5}$

1.11 Multiplying Square Roots

1) $\sqrt{6} \times \sqrt{6} = 6$

2) $\left(\sqrt{5}\right)^3 = \sqrt{5} \times \sqrt{5} \times \sqrt{5} = 5 \times \sqrt{5} = 5\sqrt{5}$

3) $\left(\sqrt{10}\right)^4 = \sqrt{10} \times \sqrt{10} \times \sqrt{10} \times \sqrt{10} = 10 \times 10 = 100$

4) $\left(\sqrt{3}\right)^3 = \sqrt{3} \times \sqrt{3} \times \sqrt{3} = 3 \times \sqrt{3} = 3\sqrt{3}$

5) $\left(\sqrt{2}\right)^5 = \sqrt{2} \times \sqrt{2} \times \sqrt{2} \times \sqrt{2} \times \sqrt{2} = 2 \times 2 \times \sqrt{2} = 4\sqrt{2}$

6) $\left(\sqrt{13}\right)^2 = \sqrt{13} \times \sqrt{13} = 13$

7) $\left(\sqrt{11}\right)^4 = \sqrt{11} \times \sqrt{11} \times \sqrt{11} \times \sqrt{11} = 11 \times 11 = 121$

8) $\left(\sqrt{5}\right)^5 = \sqrt{5} \times \sqrt{5} \times \sqrt{5} \times \sqrt{5} \times \sqrt{5} = 5 \times 5 \times \sqrt{5} = 25\sqrt{5}$

9) $\left(\sqrt{2}\right)^9 = \sqrt{2} \times \sqrt{2} \times \sqrt{2} \times \sqrt{2} \times \sqrt{2} \times \sqrt{2} \times \sqrt{2} \times \sqrt{2} \times \sqrt{2} = 2 \times 2 \times 2 \times 2 \times \sqrt{2} = 16\sqrt{2}$

10) $\left(\sqrt{3}\right)^8 = \sqrt{3} \times \sqrt{3} \times \sqrt{3} \times \sqrt{3} \times \sqrt{3} \times \sqrt{3} \times \sqrt{3} \times \sqrt{3} = 3 \times 3 \times 3 \times 3 = 9 \times 9 = 81$

11) $\sqrt{7} \times \sqrt{7} = 7$

12) $\left(\sqrt{6}\right)^6 = \sqrt{6} \times \sqrt{6} \times \sqrt{6} \times \sqrt{6} \times \sqrt{6} \times \sqrt{6} = 6 \times 6 \times 6 = 36 \times 6 = 216$

13) $\left(\sqrt{11}\right)^3 = \sqrt{11} \times \sqrt{11} \times \sqrt{11} = 11 \times \sqrt{11} = 11\sqrt{11}$

14) $\left(\sqrt{15}\right)^4 = \sqrt{15} \times \sqrt{15} \times \sqrt{15} \times \sqrt{15} = 15 \times 15 = 225$

15) $\left(\sqrt{7}\right)^5 = \sqrt{7} \times \sqrt{7} \times \sqrt{7} \times \sqrt{7} \times \sqrt{7} = 7 \times 7 \times \sqrt{7} = 49\sqrt{7}$

16) $\left(\sqrt{10}\right)^6 = \sqrt{10} \times \sqrt{10} \times \sqrt{10} \times \sqrt{10} \times \sqrt{10} \times \sqrt{10} = 10 \times 10 \times 10 = 1000$

2 Order of Operations

2.1 Addition

1) $29 + 7 = 20 + 9 + 7 = 20 + (9 + 7) = 20 + 16 = 36$

2) $13 + 13 = 10 + 3 + 10 + 3 = (10 + 10) + (3 + 3) = 20 + 6 = 26$

3) $22 + 19 = 20 + 2 + 10 + 9 = (20 + 10) + (2 + 9) = 30 + 11 = 41$

4) $44 + 31 = 40 + 4 + 30 + 1 = (40 + 30) + (4 + 1) = 70 + 5 = 75$

5) $39 + 22 = 30 + 9 + 20 + 2 = (30 + 20) + (9 + 2) = 50 + 11 = 61$

6) $16 + 73 = 10 + 6 + 70 + 3 = (10 + 70) + (6 + 3) = 80 + 9 = 89$

7) $45 + 45 = 40 + 5 + 40 + 5 = (40 + 40) + (5 + 5) = 80 + 10 = 90$

8) $68 + 76 = 60 + 8 + 70 + 6 = (60 + 70) + (8 + 6) = 130 + 14 = 144$

9) $74 + 68 = 70 + 4 + 60 + 8 = (70 + 60) + (4 + 8) = 130 + 12 = 142$

10) $87 + 79 = 80 + 7 + 70 + 9 = (80 + 70) + (7 + 9) = 150 + 16 = 166$

11) $16 + 92 = 10 + 6 + 90 + 2 = (10 + 90) + (6 + 2) = 100 + 8 = 108$

12) $99 + 99 = 90 + 9 + 90 + 9 = (90 + 90) + (9 + 9) = 180 + 18 = 198$

13) $14 + (-7) = 14 - 7 = 7$

14) $4 + (-9) = 4 - 9 = -5$

15) $-2 + (-6) = -(2 + 6) = -8$

16) $-8 + 15 = 7$

17) $7 + (-8) = 7 - 8 = -1$

18) $6 + (-7) = 6 - 7 = -1$

19) $3 + (-5) = 3 - 5 = -2$

20) $-5 + (-9) = -(5 + 9) = -14$

21) $-6 + 21 = 15$

22) $-8 + 8 = 0$

23) $-14 + 9 = -5$

24) $-9 + (-9) = -(9 + 9) = -18$

25) $-7 + (-8) = -(7 + 8) = -15$

26) $-37 + 22 = -15$

27) $36 + (-6) = 36 - 6 = 30$

28) $-11 + 5 = -6$

29) $-6 + (-8) = -(6 + 8) = -14$

30) $17 + (-19) = 17 - 19 = -2$

31) $-56 + 8 = -48$

32) $-7 + 7 = 0$

33) $-3 + (-3) = -(3 + 3) = -6$

34) $18 + (-9) = 18 - 9 = 9$

35) $-9 + (-7) = -(9 + 7) = -16$

36) $55 + (-88) = 55 - 88 = -33$

2.2 Subtraction

1) $5 - 8 = -3$

2) $-8 - 2 = -(8 + 2) = -10$

3) $4 - 9 = -5$

4) $11 - (-4) = 11 + 4 = 15$

5) $8 - 7 = 1$

6) $-7 - 3 = -(7 + 3) = -10$

7) $-5 - (-8) = -5 + 8 = 3$

8) $-10 - (-7) = -10 + 7 = -3$

9) $6 - 20 = -14$

10) $32 - 9 = 23$

11) $-7 - (-12) = -7 + 12 = 5$

12) $-8 - 13 = -(8 + 13) = -21$

13) $14 - (-5) = 14 + 5 = 19$

14) $0 - (-6) = 0 + 6 = 6$

15) $-6 - 6 = -(6 + 6) = -12$

16) $19 - (-8) = 19 + 8 = 27$

17) $99 - 33 = 66$

18) $-9 - (-9) = -9 + 9 = 0$

19) $0 - 9 = -9$

20) $-6 - (-43) = -6 + 43 = 37$

2.3 Multiplication

1) $3 \times (-9) = -27$

2) $-6 \times (-8) = 48$

3) $-4 \times 4 = -16$

4) $5 \times 6 = 30$

5) $-5 \times (-8) = 40$

6) $6 \times (-7) = -42$

7) $7 \times 8 = 56$

8) $-3 \times 9 = -27$

9) $6 \times (-9) = -54$

10) $-8 \times (-8) = 64$

11) $-4 \times 8 = -32$

12) $7 \times 9 = 63$

13) $6 \times 7 = 42$

14) $-3 \times 8 = -24$

15) $-7 \times (-9) = 63$

16) $5 \times (-7) = -35$

17) $-4 \times 7 = -28$

18) $-9 \times (-9) = 81$

19) $3 \times 9 = 27$

20) $6 \times (-5) = -30$

21) $-4 \times (-8) = 32$

22) $7 \times 8 = 56$

23) $-4 \times 6 = -24$

24) $3 \times (-5) = -15$

25) $5 \times (-9) = -45$

26) $-7 \times (-9) = 63$

27) $6 \times 9 = 54$

28) $-8 \times 9 = -72$

29) $24 \times 3 = (20 + 4) \times 3 = 20 \times 3 + 4 \times 3 = 60 + 12 = 72$

30) $7 \times 11 = 7 \times (10 + 1) = 7 \times 10 + 7 \times 1 = 70 + 7 = 77$

31) $26 \times 7 = (20 + 6) \times 7 = 20 \times 7 + 6 \times 7 = 140 + 42 = 182$

32) $5 \times 12 = 5 \times (10 + 2) + 5 \times 10 + 5 \times 2 + 50 + 10 = 60$

33) $-6 \times 14 = -6 \times (10 + 4) = -6 \times 10 - 6 \times 4 = -60 - 24 = -84$

34) $13 \times 6 = (10 + 3) \times 6 = 10 \times 6 + 3 \times 6 = 60 + 18 = 78$

35) $14 \times 7 = (10 + 4) \times 7 = 10 \times 7 + 4 \times 7 = 70 + 28 = 98$

36) $7 \times 19 = 7 \times (10 + 9) = 7 \times 10 + 7 \times 9 = 70 + 63 = 133$

37) $4 \times (-32) = 4 \times (-30 - 2) = 4 \times (-30) + 4 \times (-2) = -120 - 8 = -128$

38) $22 \times 6 = (20 + 2) \times 6 = 20 \times 6 + 2 \times 6 = 120 + 12 = 132$

39) $9 \times 41 = 9 \times (40 + 1) = 9 \times 40 + 9 \times 1 = 360 + 9 = 369$

40) $26 \times 8 = (20 + 6) \times 8 = 20 \times 8 + 6 \times 8 = 160 + 48 = 208$

41) $-5 \times (-15) = 5 \times (10 + 5) = 5 \times 10 + 5 \times 5 = 50 + 25 = 75$

42) $32 \times 6 = (30 + 2) \times 6 = 30 \times 6 + 2 \times 6 = 180 + 12 = 192$

2.4 Division

1) $28 \div 4 = 7$

2) $58 \div 9 = (54 \div 9)R(58 - 54) = 6R4$

3) $21 \div 3 = 7$

4) $36 \div 6 = 6$

5) $19 \div 5 = (15 \div 5)R(19 - 15) = 3R4$

6) $16 \div 4 = 4$

7) $56 \div 8 = 7$

8) $81 \div 9 = 9$

9) $63 \div 7 = 9$

10) $30 \div 5 = 6$

11) $41 \div 6 = (36 \div 6)R(41 - 36) = 6R5$

12) $49 \div 7 = 7$

13) $18 \div 6 = 3$

14) $54 \div 9 = 6$

15) $56 \div 7 = 8$

16) $29 \div 2 = (28 \div 2)R(29 - 28) = 14R1$

17) $32 \div 8 = 4$

18) $48 \div 6 = 8$

19) $66 \div 8 = (64 \div 8)R(66 - 64) = 8R2$

20) $63 \div 9 = 7$

21) $36 \div 4 = 9$

22) $45 \div 7 = (42 \div 7)R(45 - 42) = 6R3$

23) $72 \div 9 = 8$

24) $35 \div 5 = 7$

25) $24 \div (-6) = -4$

26) $-16 \div 8 = -2$

27) $25 \div 5 = 5$

28) $-27 \div (-3) = 9$

29) $-36 \div 6 = -6$

30) $12 \div (-4) = -3$

31) $-45 \div (-9) = 5$

32) $42 \div 7 = 6$

33) $54 \div (-6) = -9$

34) $-32 \div 8 = -4$

35) $20 \div 5 = 4$

36) $-40 \div (-8) = 5$

37) $-44 \div 4 = -11$

38) $60 \div (-10) = -6$

39) $-25 \div (-5) = 5$

40) $99 \div 9 = 11$

41) $60 \div 5 = 12$ 42) $64 \div (-8) = -8$

43) $-63 \div 7 = -9$ 44) $-36 \div (-9) = 4$

45) $50 \div (-10) = -5$ 46) $-20 \div (-4) = 5$

47) $100 \div 10 = 10$ 48) $-72 \div 8 = -9$

49) $-35 \div (-5) = 7$ 50) $48 \div (-8) = -6$

51) $-56 \div 7 = -8$ 52) $90 \div 9 = 10$

2.5 Properties of Arithmetic Operators

1) $5 - 8 = -8 + 5$ applies the commutative property of addition with $a = 5$ and $b = -8$, if you note that $5 + (-8) = 5 - 8$

2) $(4 - 2) \times 3 = 4 \times 3 - 2 \times 3$ applies the distributive property with $b = 4$, $c = -2$, and $a = 3$ (you could use the commutative property to show that this form of the distributive property is equivalent to that given in the chapter)

3) $\frac{1}{6} \times 6 = 1$ applies the inverse property of multiplication (you could use the commutative property to show that this form of the inverse property of multiplication is equivalent to that given in the chapter)

2.6 Combining Operations

1) $16 - 8 \div 2 + 2 = 16 - 4 + 2 = 12 + 2 = 14$

2) $24 \div 3 \times 2 = 8 \times 2 = 16$

3) $5 + 3 - 2 + 9 - 7 = 8 - 2 + 9 - 7 = 6 + 9 - 7 = 15 - 7 = 8$

4) $6 + 3 \times 7 + 18 \div 3 = 6 + 21 + 6 = 27 + 6 = 33$

5) $4 \times 8 - 2 \times 9 + 12 \div 4 = 32 - 18 + 3 = 14 + 3 = 17$

6) $36 \div 6 \times 2 - 8 \times 9 \div 4 = 6 \times 2 - 72 \div 4 = 12 - 18 = -6$

7) $5 + 5 \times 5 \div 5 - 5 = 5 + 25 \div 5 - 5 = 5 + 5 - 5 = 10 - 5 = 5$

8) $56 \div 7 + 3 \times 8 - 6 \times 7 = 8 + 24 - 42 = 32 - 42 = -10$

9) $3 + 6 \times 9 - 4 \times 5 + 48 \div 8 = 3 + 54 - 20 + 6 = 57 - 20 + 6 = 37 + 6 = 43$

10) $9 + 8 - 40 \div 5 \times 2 - 24 \div 6 \div 4 = 9 + 8 - 8 \times 2 - 4 \div 4 = 9 + 8 - 16 - 1 = 17 - 16 - 1$
$= 1 - 1 = 0$

11) $4 \times 7 - 5 \times 5 + 18 \div 2 - 36 \div 9 = 28 - 25 + 9 - 4 = 3 + 9 - 4 = 12 - 4 = 8$

12) $2 \div 2 + 2 \times 2 \times 2 - 2 \times 2 + 2 = 1 + 4 \times 2 - 4 + 2 = 1 + 8 - 4 + 2 = 9 - 4 + 2 = 5 + 2 = 7$

13) $81 \div 9 \div 3 + 6 \times 4 \div 3 - 42 \div 7 + 3 \times 3 = 9 \div 3 + 24 \div 3 - 6 + 9 = 3 + 8 - 6 + 9 =$
$11 - 6 + 9 = 5 + 9 = 14$

14) $9 \times 8 - 7 \times 6 - 5 \times 4 - 3 \times 2 - 1 \times 0 = 72 - 42 - 20 - 6 - 0 = 30 - 20 - 6 = 10 - 6 = 4$

15) $7 + 3 \times 5 + 6 \times 6 \div 4 - 2 \times 3 \times 4 + 8 = 7 + 15 + 36 \div 4 - 6 \times 4 + 8 = 7 + 15 + 9 - 24 + 8$
$= 22 + 9 - 24 + 8 = 31 - 24 + 8 = 7 + 8 = 15$

2.7 Operations with Exponents

1) $15 - 3^2 + 2^4 = 15 - 9 + 16 = 6 + 16 = 22$

2) $8 - 4^2 \div 2^2 = 8 - 16 \div 4 = 8 - 4 = 4$

3) $6 + 2 \times 5^2 - 7^2 = 6 + 2 \times 25 - 49 = 6 + 50 - 49 = 56 - 49 = 7$

4) $9^2 - 3 \times 5^2 + 24 \div 6 = 81 - 3 \times 25 + 4 = 81 - 75 + 4 = 6 + 4 = 10$

5) $4^3 + 6^2 - 2^2 \times 5^2 = 64 + 36 - 4 \times 25 = 64 + 36 - 100 = 100 - 100 = 0$

6) $\sqrt[5]{32} \times 3 + 8 = 2 \times 3 + 8 = 6 + 8 = 14$

7) $8^2 \div 4^2 \times 3^2 \div 2^2 = 64 \div 16 \times 9 \div 4 = 4 \times 9 \div 4 = 36 \div 4 = 9$

2.8 Operations with Parentheses and Exponents

1) $9 - (3 + 5) \div 2 = 9 - 8 \div 2 = 9 - 4 = 5$

2) $(12 - 5) \times (7 - 4) = 7 \times 3 = 21$

3) $3 \times (9 - 2^2) - 3^2 = 3 \times (9 - 4) - 3^2 = 3 \times 5 - 3^2 = 3 \times 5 - 9 = 15 - 9 = 6$

4) $(4 + 2 \times 3) \div 2 + 3 = (4 + 6) \div 2 + 3 = 10 \div 2 + 3 = 5 + 3 = 8$

5) $(4 \times 6 - 2 \times 9)^2 = (24 - 18)^2 = 6^2 = 36$

6) $(-3)^2 \div (-3) = 9 \div (-3) = -3$

7) $(5^2 - 4^2)^2 = (25 - 16)^2 = 9^2 = 81$

8) $(3 - 8) \times 4 + 4^2 = -5 \times 4 + 4^2 = -5 \times 4 + 16 = -20 + 16 = -4$

9) $-3 - 4 \times (1 + 2) = -3 - 4 \times 3 = -3 - 12 = -15$

10) $-2 - 2^2 - (-2)^2 = -2 - 4 - (4) = -2 - 4 - 4 = -6 - 4 = -10$

11) $(7 - 5 \times 2)^3 - (4 - 2 \times 3)^2 = (7 - 10)^3 - (4 - 6)^2 = (-3)^3 - (-2)^2 = -27 - 4 = -31$

12) $(2 + 3) \times (9 - 5) \div (8 - 6) = 5 \times 4 \div 2 = 20 \div 2 = 10$

13) $2^{9 - 3 \times (6 - 4)} = 2^{9 - 3 \times 2} = 2^{9 - 6} = 2^3 = 8$

14) $\sqrt[3]{64 - 8 \times (11 - 5) + 11} = \sqrt[3]{64 - 8 \times 6 + 11} = \sqrt[3]{64 - 48 + 11} = \sqrt[3]{27} = 3$

2.9 Why the Order of Operations Matters

1) With: $24 \div (4 + 2) = 24 \div 6 = 4$
Without: $24 \div 4 + 2 = 6 + 2 = 8$
2) With: $(6 + 8) \div 2 = 14 \div 2 = 7$
Without: $6 + 8 \div 2 = 6 + 4 = 10$
3) With: $3 \times (3 + 3) = 3 \times 6 = 18$
Without: $3 \times 3 + 3 = 9 + 3 = 12$
4) With: $(15 - 6) \times 3 = 9 \times 3 = 27$
Without: $15 - 6 \times 3 = 15 - 18 = -3$
5) With: $(4 \times 9) \div (3 \times 4) = 36 \div 12 = 3$
Without: $4 \times 9 \div 3 \times 4 = 36 \div 3 \times 4 = 12 \times 4 = 48$

2.10 Parentheses Challenge

1) $3 \times (4 + 2) \times 5 = 3 \times 6 \times 5 = 18 \times 5 = 90$
2) $72 \div (4 \times 2) - 6 = 72 \div 8 - 6 = 9 - 6 = 3$
3) $4 + 4 \times (4 - 4) = 4 + 4 \times 0 = 4 + 0 = 4$
4) $(5 + 3) \times (9 - 2) = 8 \times 7 = 56$
5) $(5 \times 6 - 3 \times 8 + 8) \div 2 = (30 - 24 + 8) \div 2 = (6 + 8) \div 2 = 14 \div 2 = 7$
6) $(13 - 4) \times (2 + 3) - 7 \times 6 = 9 \times 5 - 7 \times 6 = 45 - 42 = 3$

3 Fractions

3.1 Reducing Fractions

1) $\frac{8}{12} = \frac{8 \div 4}{12 \div 4} = \frac{2}{3}$

2) $\frac{20}{40} = \frac{20 \div 20}{40 \div 20} = \frac{1}{2}$

3) $\frac{9}{21} = \frac{9 \div 3}{21 \div 3} = \frac{3}{7}$

4) $\frac{25}{10} = \frac{25 \div 5}{10 \div 5} = \frac{5}{2}$

5) $\frac{10}{12} = \frac{10 \div 2}{12 \div 2} = \frac{5}{6}$

6) $\frac{8}{32} = \frac{8 \div 8}{32 \div 8} = \frac{1}{4}$

7) $\frac{18}{24} = \frac{18 \div 6}{24 \div 6} = \frac{3}{4}$

8) $\frac{16}{28} = \frac{16 \div 4}{28 \div 4} = \frac{4}{7}$

9) $\frac{32}{18} = \frac{32 \div 2}{18 \div 2} = \frac{16}{9}$

10) $\frac{9}{15} = \frac{9 \div 3}{15 \div 3} = \frac{3}{5}$

11) $\frac{14}{49} = \frac{14 \div 7}{49 \div 7} = \frac{2}{7}$

12) $\frac{45}{72} = \frac{45 \div 9}{72 \div 9} = \frac{5}{8}$

13) $\frac{10}{25} = \frac{10 \div 5}{25 \div 5} = \frac{2}{5}$

14) $\frac{36}{8} = \frac{36 \div 4}{8 \div 4} = \frac{9}{2}$

15) $\frac{42}{7} = \frac{42 \div 7}{7 \div 7} = \frac{6}{1} = 6$

16) $\frac{24}{16} = \frac{24 \div 8}{16 \div 8} = \frac{3}{2}$

17) $\frac{45}{18} = \frac{45 \div 9}{18 \div 9} = \frac{5}{2}$

18) $\frac{32}{48} = \frac{32 \div 16}{48 \div 16} = \frac{2}{3}$

3.2 Mixed Numbers

1) $3\frac{4}{9} = \frac{3 \times 9 + 4}{9} = \frac{31}{9}$

2) $6\frac{3}{4} = \frac{6 \times 4 + 3}{4} = \frac{27}{4}$

3) $2\frac{6}{7} = \frac{2 \times 7 + 6}{7} = \frac{20}{7}$

4) $5\frac{3}{5} = \frac{5 \times 5 + 3}{5} = \frac{28}{5}$

5) $7\frac{2}{5} = \frac{7 \times 5 + 2}{5} = \frac{37}{5}$

6) $9\frac{1}{6} = \frac{9 \times 6 + 1}{6} = \frac{55}{6}$

7) $4\frac{7}{8} = \frac{4 \times 8 + 7}{8} = \frac{39}{8}$

8) $5\frac{5}{9} = \frac{5 \times 9 + 5}{9} = \frac{50}{9}$

9) $8\frac{3}{5} = \frac{8 \times 5 + 3}{5} = \frac{43}{5}$

10) $2\frac{7}{8} = \frac{2 \times 8 + 7}{8} = \frac{23}{8}$

11) $6\frac{4}{7} = \frac{6 \times 7 + 4}{7} = \frac{46}{7}$

12) $7\frac{2}{9} = \frac{7 \times 9 + 2}{9} = \frac{65}{9}$

13) $4\frac{4}{5} = \frac{4 \times 5 + 4}{5} = \frac{24}{5}$

14) $8\frac{5}{6} = \frac{8 \times 6 + 5}{6} = \frac{53}{6}$

15) $5\frac{6}{7} = \frac{5 \times 7 + 6}{7} = \frac{41}{7}$

16) $3\frac{3}{8} = \frac{3 \times 8 + 3}{8} = \frac{27}{8}$

17) $9\frac{5}{6} = \frac{9 \times 6 + 5}{6} = \frac{59}{6}$

18) $7\frac{7}{9} = \frac{7 \times 9 + 7}{9} = \frac{70}{9}$

19) $6\frac{7}{8} = \frac{6 \times 8 + 7}{8} = \frac{55}{8}$

20) $9\frac{8}{9} = \frac{9 \times 9 + 8}{9} = \frac{89}{9}$

21) $\frac{7}{2} = 7 \div 2 = 3R1 = 3\frac{1}{2}$

22) $\frac{20}{3} = 20 \div 3 = 6R2 = 6\frac{2}{3}$

23) $\frac{23}{6} = 23 \div 6 = 3R5 = 3\frac{5}{6}$

24) $\frac{19}{7} = 19 \div 7 = 2R5 = 2\frac{5}{7}$

25) $\frac{35}{8} = 35 \div 8 = 4R3 = 4\frac{3}{8}$

26) $\frac{29}{5} = 29 \div 5 = 5R4 = 5\frac{4}{5}$

27) $\frac{34}{9} = 34 \div 9 = 3R7 = 3\frac{7}{9}$

28) $\frac{44}{7} = 44 \div 7 = 6R2 = 6\frac{2}{7}$

29) $\frac{39}{4} = 39 \div 4 = 9R3 = 9\frac{3}{4}$

30) $\frac{45}{8} = 45 \div 8 = 5R5 = 5\frac{5}{8}$

31) $\frac{29}{6} = 29 \div 6 = 4R5 = 4\frac{5}{6}$

32) $\frac{26}{3} = 26 \div 3 = 8R2 = 8\frac{2}{3}$

33) $\frac{49}{6} = 49 \div 6 = 8R1 = 8\frac{1}{6}$

34) $\frac{31}{4} = 31 \div 4 = 7R3 = 7\frac{3}{4}$

35) $\frac{68}{7} = 68 \div 7 = 9R5 = 9\frac{5}{7}$

36) $\frac{16}{9} = 16 \div 9 = 1R7 = 1\frac{7}{9}$

37) $\frac{71}{8} = 71 \div 8 = 8R7 = 8\frac{7}{8}$

38) $\frac{52}{7} = 52 \div 7 = 7R3 = 7\frac{3}{7}$

39) $\frac{41}{6} = 41 \div 6 = 6R5 = 6\frac{5}{6}$

40) $\frac{89}{9} = 89 \div 9 = 9R8 = 9\frac{8}{9}$

3.3 Adding and Subtracting Fractions

1) $\frac{3}{4} + \frac{1}{6} = \frac{3\times3}{4\times3} + \frac{1\times2}{6\times2} = \frac{9}{12} + \frac{2}{12} = \frac{9+2}{12} = \frac{11}{12}$

2) $\frac{5}{6} - \frac{3}{8} = \frac{5\times4}{6\times4} - \frac{3\times3}{8\times3} = \frac{20}{24} - \frac{9}{24} = \frac{20-9}{24} = \frac{11}{24}$

3) $\frac{2}{3} + \frac{3}{5} = \frac{2\times5}{3\times5} + \frac{3\times3}{5\times3} = \frac{10}{15} + \frac{9}{15} = \frac{10+9}{15} = \frac{19}{15}$ or $1\frac{4}{15}$

4) $\frac{7}{6} - \frac{2}{3} = \frac{7\times1}{6\times1} - \frac{2\times2}{3\times2} = \frac{7}{6} - \frac{4}{6} = \frac{7-4}{6} = \frac{3}{6} = \frac{3\div3}{6\div3} = \frac{1}{2}$

5) $\frac{7}{10} + \frac{4}{15} = \frac{7\times3}{10\times3} + \frac{4\times2}{15\times2} = \frac{21}{30} + \frac{8}{30} = \frac{21+8}{30} = \frac{29}{30}$

6) $\frac{11}{12} - \frac{1}{4} = \frac{11\times1}{12\times1} - \frac{1\times3}{4\times3} = \frac{11}{12} - \frac{3}{12} = \frac{11-3}{12} = \frac{8}{12} = \frac{8\div4}{12\div4} = \frac{2}{3}$

7) $\frac{3}{2} + \frac{2}{3} = \frac{3\times3}{2\times3} + \frac{2\times2}{3\times2} = \frac{9}{6} + \frac{4}{6} = \frac{9+4}{6} = \frac{13}{6}$ or $2\frac{1}{6}$

8) $\frac{9}{8} - \frac{5}{12} = \frac{9\times3}{8\times3} - \frac{5\times2}{12\times2} = \frac{27}{24} - \frac{10}{24} = \frac{27-10}{24} = \frac{17}{24}$

9) $\frac{7}{12} + \frac{5}{18} = \frac{7\times3}{12\times3} + \frac{5\times2}{18\times2} = \frac{21}{36} + \frac{10}{36} = \frac{21+10}{36} = \frac{31}{36}$

10) $\frac{8}{15} - \frac{2}{9} = \frac{8\times3}{15\times3} - \frac{2\times5}{9\times5} = \frac{24}{45} - \frac{10}{45} = \frac{24-10}{45} = \frac{14}{45}$

3.4 Multiplying Fractions

1) $\frac{3}{4} \times \frac{5}{7} = \frac{3 \times 5}{4 \times 7} = \frac{15}{28}$

2) $\frac{1}{2} \times \frac{4}{7} = \frac{1 \times 4}{2 \times 7} = \frac{4}{14} = \frac{4 \div 2}{14 \div 2} = \frac{2}{7}$

3) $\frac{4}{3} \times \frac{5}{8} = \frac{4 \times 5}{3 \times 8} = \frac{20}{24} = \frac{20 \div 4}{24 \div 4} = \frac{5}{6}$

4) $\frac{3}{4} \times \frac{3}{4} = \frac{3 \times 3}{4 \times 4} = \frac{9}{16}$

5) $\frac{4}{3} \times \frac{3}{4} = \frac{4 \times 3}{3 \times 4} = \frac{12}{12} = \frac{12 \div 12}{12 \div 12} = \frac{1}{1} = 1$

6) $\frac{9}{5} \times \frac{10}{3} = \frac{9 \times 10}{5 \times 3} = \frac{90}{15} = \frac{90 \div 15}{15 \div 15} = \frac{6}{1} = 6$

7) $\frac{1}{2} \times \frac{1}{3} = \frac{1 \times 1}{2 \times 3} = \frac{1}{6}$

8) $\frac{7}{12} \times \frac{8}{3} = \frac{7 \times 8}{12 \times 3} = \frac{56}{36} = \frac{56 \div 4}{36 \div 4} = \frac{14}{9}$ or $1\frac{5}{9}$

9) $\frac{15}{16} \times \frac{4}{5} = \frac{15 \times 4}{16 \times 5} = \frac{60}{80} = \frac{60 \div 20}{80 \div 20} = \frac{3}{4}$

10) $\frac{3}{5} \times \frac{25}{12} = \frac{3 \times 25}{5 \times 12} = \frac{75}{60} = \frac{75 \div 15}{60 \div 15} = \frac{5}{4}$ or $1\frac{1}{4}$

3.5 Reciprocals

1) $\left(\frac{4}{3}\right)^{-1} = \frac{3}{4}$

2) $\left(\frac{5}{7}\right)^{-1} = \frac{7}{5}$

3) $\left(\frac{3}{5}\right)^{-1} = \frac{5}{3}$

4) $\left(\frac{9}{10}\right)^{-1} = \frac{10}{9}$

5) $\left(\frac{8}{7}\right)^{-1} = \frac{7}{8}$

6) $\left(\frac{6}{11}\right)^{-1} = \frac{11}{6}$

7) $\left(\frac{1}{8}\right)^{-1} = \frac{8}{1} = 8$

8) $\left(\frac{6}{7}\right)^{-1} = \frac{7}{6}$

9) $\left(\frac{4}{5}\right)^{-1} = \frac{5}{4}$

10) $\left(\frac{12}{7}\right)^{-1} = \frac{7}{12}$

11) $\left(\frac{10}{9}\right)^{-1} = \frac{9}{10}$

12) $\left(\frac{5}{6}\right)^{-1} = \frac{6}{5}$

13) $\left(\frac{9}{13}\right)^{-1} = \frac{13}{9}$

14) $9^{-1} = \frac{1}{9}$

15) $\left(\frac{14}{11}\right)^{-1} = \frac{11}{14}$

16) $\left(\frac{15}{8}\right)^{-1} = \frac{8}{15}$

17) $\left(\frac{16}{13}\right)^{-1} = \frac{13}{16}$

18) $\left(\frac{11}{12}\right)^{-1} = \frac{12}{11}$

3.6 Dividing Fractions

1) $\dfrac{3}{5} \div \dfrac{4}{7} = \dfrac{3}{5} \times \dfrac{7}{4} = \dfrac{3 \times 7}{5 \times 4} = \dfrac{21}{20}$ or $1\dfrac{1}{20}$

2) $\dfrac{5}{4} \div \dfrac{1}{2} = \dfrac{5}{4} \times \dfrac{2}{1} = \dfrac{5 \times 2}{4 \times 1} = \dfrac{10}{4} = \dfrac{10 \div 2}{4 \div 2} = \dfrac{5}{2}$ or $2\dfrac{1}{2}$

3) $\dfrac{8}{9} \div \dfrac{2}{3} = \dfrac{8}{9} \times \dfrac{3}{2} = \dfrac{8 \times 3}{9 \times 2} = \dfrac{24}{18} = \dfrac{24 \div 6}{18 \div 6} = \dfrac{4}{3}$ or $1\dfrac{1}{3}$

4) $\dfrac{1}{3} \div \dfrac{1}{4} = \dfrac{1}{3} \times \dfrac{4}{1} = \dfrac{1 \times 4}{3 \times 1} = \dfrac{4}{3}$ or $1\dfrac{1}{3}$

5) $\dfrac{2}{5} \div \dfrac{2}{5} = \dfrac{2}{5} \times \dfrac{5}{2} = \dfrac{2 \times 5}{5 \times 2} = \dfrac{10}{10} = \dfrac{10 \div 10}{10 \div 10} = \dfrac{1}{1} = 1$

6) $\dfrac{5}{2} \div \dfrac{2}{5} = \dfrac{5}{2} \times \dfrac{5}{2} = \dfrac{5 \times 5}{2 \times 2} = \dfrac{25}{4}$ or $6\dfrac{1}{4}$

7) $\dfrac{3}{7} \div \dfrac{6}{5} = \dfrac{3}{7} \times \dfrac{5}{6} = \dfrac{3 \times 5}{7 \times 6} = \dfrac{15}{42} = \dfrac{15 \div 3}{42 \div 3} = \dfrac{5}{14}$

8) $\dfrac{5}{12} \div \dfrac{3}{4} = \dfrac{5}{12} \times \dfrac{4}{3} = \dfrac{5 \times 4}{12 \times 3} = \dfrac{20}{36} = \dfrac{20 \div 4}{36 \div 4} = \dfrac{5}{9}$

9) $\dfrac{21}{20} \div \dfrac{7}{5} = \dfrac{21}{20} \times \dfrac{5}{7} = \dfrac{21 \times 5}{20 \times 7} = \dfrac{105}{140} = \dfrac{105 \div 35}{140 \div 35} = \dfrac{3}{4}$

10) $\dfrac{11}{6} \div \dfrac{2}{3} = \dfrac{11}{6} \times \dfrac{3}{2} = \dfrac{11 \times 3}{6 \times 2} = \dfrac{33}{12} = \dfrac{33 \div 3}{12 \div 3} = \dfrac{11}{4}$ or $2\dfrac{3}{4}$

11) $\dfrac{9}{10} \div 3 = \dfrac{9}{10} \times \dfrac{1}{3} = \dfrac{9 \times 1}{10 \times 3} = \dfrac{9}{30} = \dfrac{9 \div 3}{30 \div 3} = \dfrac{3}{10}$

3.7 Powers of Fractions

1) $\left(\dfrac{4}{7}\right)^{3} = \dfrac{4^{3}}{7^{3}} = \dfrac{4 \times 4 \times 4}{7 \times 7 \times 7} = \dfrac{16 \times 4}{49 \times 7} = \dfrac{64}{343}$

2) $\left(\dfrac{4}{3}\right)^{4} = \dfrac{4^{4}}{3^{4}} = \dfrac{4 \times 4 \times 4 \times 4}{3 \times 3 \times 3 \times 3} = \dfrac{16 \times 16}{9 \times 9} = \dfrac{256}{81}$

3) $\left(\dfrac{5}{6}\right)^{2} = \dfrac{5^{2}}{6^{2}} = \dfrac{5 \times 5}{6 \times 6} = \dfrac{25}{36}$

4) $\left(\dfrac{8}{9}\right)^{1} = \dfrac{8^{1}}{9^{1}} = \dfrac{8}{9}$

5) $\left(\dfrac{3}{2}\right)^{5} = \dfrac{3^{5}}{2^{5}} = \dfrac{3 \times 3 \times 3 \times 3 \times 3}{2 \times 2 \times 2 \times 2 \times 2} = \dfrac{9 \times 9 \times 3}{4 \times 4 \times 2} = \dfrac{81 \times 3}{16 \times 2} = \dfrac{243}{32}$

6) $\left(\dfrac{3}{5}\right)^{4} = \dfrac{3^{4}}{5^{4}} = \dfrac{3 \times 3 \times 3 \times 3}{5 \times 5 \times 5 \times 5} = \dfrac{9 \times 9}{25 \times 25} = \dfrac{81}{625}$

7) $\left(\dfrac{8}{7}\right)^{3} = \dfrac{8^{3}}{7^{3}} = \dfrac{8 \times 8 \times 8}{7 \times 7 \times 7} = \dfrac{64 \times 8}{49 \times 7} = \dfrac{512}{343}$

8) $\left(\dfrac{2}{7}\right)^{0} = \dfrac{2^{0}}{7^{0}} = \dfrac{1}{1} = 1$

9) $\left(\frac{6}{5}\right)^4 = \frac{6^4}{5^4} = \frac{6\times6\times6\times6}{5\times5\times5\times5} = \frac{36\times36}{25\times25} = \frac{1296}{625}$

10) $\left(\frac{8}{9}\right)^3 = \frac{8^3}{9^3} = \frac{8\times8\times8}{9\times9\times9} = \frac{64\times8}{81\times9} = \frac{512}{729}$

3.8 Negative Powers of Fractions

1) $\left(\frac{5}{8}\right)^{-3} = \left(\frac{8}{5}\right)^3 = \frac{8^3}{5^3} = \frac{8\times8\times8}{5\times5\times5} = \frac{64\times8}{25\times5} = \frac{512}{125}$

2) $\left(\frac{4}{7}\right)^{-2} = \left(\frac{7}{4}\right)^2 = \frac{7^2}{4^2} = \frac{7\times7}{4\times4} = \frac{49}{16}$

3) $\left(\frac{6}{5}\right)^{-4} = \left(\frac{5}{6}\right)^4 = \frac{5^4}{6^4} = \frac{5\times5\times5\times5}{6\times6\times6\times6} = \frac{25\times25}{36\times36} = \frac{625}{1296}$

4) $\left(\frac{8}{9}\right)^{-1} = \left(\frac{9}{8}\right)^1 = \frac{9^1}{8^1} = \frac{9}{8}$

5) $\left(\frac{6}{7}\right)^{-3} = \left(\frac{7}{6}\right)^3 = \frac{7^3}{6^3} = \frac{7\times7\times7}{6\times6\times6} = \frac{49\times7}{36\times6} = \frac{343}{216}$

6) $\left(\frac{2}{3}\right)^{-5} = \left(\frac{3}{2}\right)^5 = \frac{3^5}{2^5} = \frac{3\times3\times3\times3\times3}{2\times2\times2\times2\times2} = \frac{9\times9\times3}{4\times4\times2} = \frac{81\times3}{16\times2} = \frac{243}{32}$

7) $\left(\frac{5}{6}\right)^{-3} = \left(\frac{6}{5}\right)^3 = \frac{6^3}{5^3} = \frac{6\times6\times6}{5\times5\times5} = \frac{36\times6}{25\times5} = \frac{216}{125}$

8) $\left(\frac{9}{2}\right)^{-4} = \left(\frac{2}{9}\right)^4 = \frac{2^4}{9^4} = \frac{2\times2\times2\times2}{9\times9\times9\times9} = \frac{4\times4}{81\times81} = \frac{16}{6561}$

9) $\left(\frac{7}{3}\right)^{-0} = \left(\frac{3}{7}\right)^0 = \frac{3^0}{7^0} = \frac{1}{1} = 1$

10) $\left(\frac{3}{10}\right)^{-5} = \left(\frac{10}{3}\right)^5 = \frac{10^5}{3^5} = \frac{10\times10\times10\times10\times10}{3\times3\times3\times3\times3} = \frac{100\times100\times10}{9\times9\times3} = \frac{10,000\times10}{81\times3} = \frac{100,000}{243}$

3.9 Fractional Powers

1) $27^{4/3} = \left(\sqrt[3]{27}\right)^4 = 3^4 = 3\times3\times3\times3 = 9\times9 = 81$

2) $64^{7/6} = \left(\sqrt[6]{64}\right)^7 = (\pm2)^7 = \pm2\times2\times2\times2\times2\times2\times2 = \pm4\times4\times4\times2 = \pm16\times8 = \pm128$

3) $81^{-3/4} = \left(\frac{1}{81}\right)^{3/4} = \left(\frac{1}{\sqrt[4]{81}}\right)^3 = \left(\pm\frac{1}{3}\right)^3 = \pm\frac{1}{3^3} = \pm\frac{1}{3\times3\times3} = \pm\frac{1}{9\times3} = \pm\frac{1}{27}$

4) $4^{-1/2} = \left(\frac{1}{4}\right)^{1/2} = \left(\frac{1}{\sqrt{4}}\right)^1 = \left(\pm\frac{1}{2}\right)^1 = \pm\frac{1}{2}$

5) $100,000^{6/5} = \left(\sqrt[5]{100,000}\right)^6 = 10^6 = 10\times10\times10\times10\times10\times10 = 1000\times1000 = 1,000,000$

6) $125^{-1/3} = \left(\frac{1}{125}\right)^{1/3} = \left(\frac{1}{\sqrt[3]{125}}\right)^1 = \left(\frac{1}{5}\right)^1 = \frac{1}{5}$

7) $216^{2/3} = \left(\sqrt[3]{216}\right)^2 = 6^2 = 6 \times 6 = 36$

8) $1024^{-4/5} = \left(\frac{1}{1024}\right)^{4/5} = \left(\frac{1}{\sqrt[5]{1024}}\right)^4 = \left(\frac{1}{4}\right)^4 = \frac{1}{4^4} = \frac{1}{4 \times 4 \times 4 \times 4} = \frac{1}{16 \times 16} = \frac{1}{256}$

9) $100^{-5/2} = \left(\frac{1}{100}\right)^{5/2} = \left(\frac{1}{\sqrt{100}}\right)^5 = \left(\pm\frac{1}{10}\right)^5 = \pm\frac{1}{10^5} = \pm\frac{1}{10 \times 10 \times 10 \times 10 \times 10} = \pm\frac{1}{100 \times 1000} = \pm\frac{1}{100,000}$

10) $64^{-5/3} = \left(\frac{1}{64}\right)^{5/3} = \left(\frac{1}{\sqrt[3]{64}}\right)^5 = \left(\frac{1}{4}\right)^5 = \frac{1}{4^5} = \frac{1}{4 \times 4 \times 4 \times 4 \times 4} = \frac{1}{16 \times 16 \times 4} = \frac{1}{256 \times 4} = \frac{1}{1024}$

11) $36^{3/2} = \left(\sqrt{36}\right)^3 = (\pm 6)^3 = \pm 6 \times 6 \times 6 = \pm 36 \times 6 = \pm 216$

12) $\left(\frac{625}{256}\right)^{1/4} = \left(\sqrt[4]{\frac{625}{256}}\right)^1 = \left(\frac{\sqrt[4]{625}}{\sqrt[4]{256}}\right)^1 = \left(\pm\frac{5}{4}\right)^1 = \pm\frac{5}{4}$

13) $\left(\frac{49}{81}\right)^{-3/2} = \left(\frac{81}{49}\right)^{3/2} = \left(\sqrt{\frac{81}{49}}\right)^3 = \left(\frac{\sqrt{81}}{\sqrt{49}}\right)^3 = \left(\pm\frac{9}{7}\right)^3 = \pm\frac{9^3}{7^3} = \pm\frac{9 \times 9 \times 9}{7 \times 7 \times 7} = \pm\frac{81 \times 9}{49 \times 7} = \pm\frac{729}{343}$

14) $\left(\frac{1}{343}\right)^{-1/3} = (343)^{1/3} = \left(\sqrt[3]{343}\right)^1 = 7^1 = 7$

15) $\left(\frac{32}{243}\right)^{4/5} = \left(\sqrt[5]{\frac{32}{243}}\right)^4 = \left(\frac{\sqrt[5]{32}}{\sqrt[5]{243}}\right)^4 = \left(\frac{2}{3}\right)^4 = \frac{2^4}{3^4} = \frac{2 \times 2 \times 2 \times 2}{3 \times 3 \times 3 \times 3} = \frac{4 \times 4}{9 \times 9} = \frac{16}{81}$

16) $\left(\frac{729}{1,000,000}\right)^{-5/6} = \left(\frac{1,000,000}{729}\right)^{5/6} = \left(\sqrt[6]{\frac{1,000,000}{729}}\right)^5 = \left(\frac{\sqrt[6]{1,000,000}}{\sqrt[6]{729}}\right)^5 = \left(\pm\frac{10}{3}\right)^5 = \pm\frac{10^5}{3^5}$

$= \pm\frac{10 \times 10 \times 10 \times 10 \times 10}{3 \times 3 \times 3 \times 3 \times 3} = \pm\frac{100 \times 1000}{9 \times 9 \times 3} = \pm\frac{100,000}{81 \times 3} = \pm\frac{100,000}{243}$

3.10 Rationalize the Denominator

1) $\frac{1}{\sqrt{3}} = \frac{1 \times \sqrt{3}}{\sqrt{3} \times \sqrt{3}} = \frac{\sqrt{3}}{3}$

2) $\frac{6}{\sqrt{2}} = \frac{6 \times \sqrt{2}}{\sqrt{2} \times \sqrt{2}} = \frac{6 \times \sqrt{2}}{2} = \frac{6 \times \sqrt{2} \div 2}{2 \div 2} = \frac{3 \times \sqrt{2}}{1} = 3 \times \sqrt{2}$

3) $\frac{7}{\sqrt{7}} = \frac{7 \times \sqrt{7}}{\sqrt{7} \times \sqrt{7}} = \frac{7 \times \sqrt{7}}{7} = \frac{7 \times \sqrt{7} \div 7}{7 \div 7} = \frac{\sqrt{7}}{1} = \sqrt{7}$

4) $\frac{50}{\sqrt{10}} = \frac{50 \times \sqrt{10}}{\sqrt{10} \times \sqrt{10}} = \frac{50 \times \sqrt{10}}{10} = \frac{50 \times \sqrt{10} \div 10}{10 \div 10} = \frac{5 \times \sqrt{10}}{1} = 5 \times \sqrt{10}$

5) $\frac{1}{2 \times \sqrt{6}} = \frac{1 \times \sqrt{6}}{2 \times \sqrt{6} \times \sqrt{6}} = \frac{\sqrt{6}}{2 \times 6} = \frac{\sqrt{6}}{12}$

4 Decimals

4.1 Place Value

1) 0.24: tenths = 2, hundredths = 4

2) 1.732: tenths = 7, hundredths = 3, thousandths = 2

3) 9.7531: tenths = 7, hundredths = 5, thousandths = 3, ten thousandths = 1

4) 3.14159: tenths = first 1, hundredths = 4, thousandths = second 1, ten thousandths = 5, hundred thousandths = 9

4.2 Powers of Ten

1) $10^2 = 10 \times 10 = 100$ (2 zeroes after)

2) $10^{-4} = \frac{1}{10^4} = 0.0001$ (3 zeroes in between since $4 - 1 = 3$)

3) $10^7 = 10 \times 10 \times 10 \times 10 \times 10 \times 10 \times 10 = 10{,}000{,}000$ (7 zeroes after)

4) $10^{-2} = \frac{1}{10^2} = 0.01$ (1 zero in between since $2 - 1 = 1$)

5) $10^4 = 10 \times 10 \times 10 \times 10 = 10{,}000$ (4 zeroes after)

6) $10^{-1} = \frac{1}{10^1} = 0.1$ (*no* zeroes in between since $1 - 1 = 0$)

7) $10^6 = 10 \times 10 \times 10 \times 10 \times 10 \times 10 = 1{,}000{,}000$ (6 zeroes after)

8) $10^1 = 10 \times 1 = 10$ (1 zero after)

9) $10^{-8} = \frac{1}{10^8} = 0.00000001$ (7 zeroes in between since $8 - 1 = 7$)

10) $10^0 = 1$ (no zeroes after)

11) $10^{-6} = \frac{1}{10^6} = 0.000001$ (5 zeroes in between since $6 - 1 = 5$)

12) $10^{-9} = \frac{1}{10^9} = 0.000000001$ (8 zeroes in between since $9 - 1 = 8$)

4.3 Scientific Notation

1) $7295 = 7.295 \times 10^3$ (left 3)

2) $0.000168 = 1.68 \times 10^{-4}$ (right 4)

3) $0.003876 = 3.876 \times 10^{-3}$ (right 3)

4) $638 = 6.38 \times 10^2$ (left 2)

5) $989.898 = 9.89898 \times 10^2$ (left 2)

6) $0.00000687 = 6.87 \times 10^{-6}$ (right 6)

7) $0.044 = 4.4 \times 10^{-2}$ (right 2)

8) $8479.2 = 8.4792 \times 10^3$ (left 3)

9) $412{,}567.43 = 4.1256743 \times 10^5$ (left 5)

10) $0.0000888 = 8.88 \times 10^{-5}$ (right 5)

11) $0.21 = 2.1 \times 10^{-1}$ (right 1)

12) $6{,}642{,}155 = 6.642155 \times 10^6$ (left 6)

13) $615{,}842.392 = 6.15842392 \times 10^5$ (left 5)

14) $0.00000001 = 1 \times 10^{-8}$ (right 8)

15) $0.0254123 = 2.54123 \times 10^{-2}$ (right 2)

16) $99{,}999.99 = 9.999999 \times 10^4$ (left 4)

4.4 Converting Decimals to Fractions

1) $0.6 = \dfrac{6}{10} = \dfrac{6 \div 2}{10 \div 2} = \dfrac{3}{5}$

2) $0.49 = \dfrac{49}{100}$

3) $1.5 = \dfrac{15}{10} = \dfrac{15 \div 5}{10 \div 5} = \dfrac{3}{2}$

4) $0.75 = \dfrac{75}{100} = \dfrac{75 \div 25}{100 \div 25} = \dfrac{3}{4}$

5) $0.36 = \dfrac{36}{100} = \dfrac{36 \div 4}{100 \div 4} = \dfrac{9}{25}$

6) $0.217 = \dfrac{217}{1000}$

7) $0.375 = \dfrac{375}{1000} = \dfrac{375 \div 125}{1000 \div 125} = \dfrac{3}{8}$

8) $1.23 = \dfrac{123}{100}$

9) $3.2 = \dfrac{32}{10} = \dfrac{32 \div 2}{10 \div 2} = \dfrac{16}{5}$

10) $0.05 = \dfrac{5}{100} = \dfrac{5 \div 5}{100 \div 5} = \dfrac{1}{20}$

11) $0.5 = \dfrac{5}{10} = \dfrac{5 \div 5}{10 \div 5} = \dfrac{1}{2}$

12) $1.25 = \dfrac{125}{100} = \dfrac{125 \div 25}{100 \div 25} = \dfrac{5}{4}$

13) $1.64 = \dfrac{164}{100} = \dfrac{164 \div 4}{100 \div 4} = \dfrac{41}{25}$

14) $0.004 = \dfrac{4}{1000} = \dfrac{4 \div 4}{1000 \div 4} = \dfrac{1}{250}$

4.5 Converting Fractions to Decimals

1) $\dfrac{3}{4} = \dfrac{3 \times 25}{4 \times 25} = \dfrac{75}{100} = 0.75$

2) $\dfrac{3}{2} = \dfrac{3 \times 5}{2 \times 5} = \dfrac{15}{10} = 1.5$

3) $\dfrac{2}{5} = \dfrac{2 \times 2}{5 \times 2} = \dfrac{4}{10} = 0.4$

4) $\dfrac{5}{8} = \dfrac{5 \times 125}{8 \times 125} = \dfrac{625}{1000} = 0.625$

5) $\dfrac{17}{20} = \dfrac{17 \times 5}{20 \times 5} = \dfrac{85}{100} = 0.85$

6) $\dfrac{9}{40} = \dfrac{9 \times 25}{40 \times 25} = \dfrac{225}{1000} = 0.225$

7) $\dfrac{4}{25} = \dfrac{4 \times 4}{25 \times 4} = \dfrac{16}{100} = 0.16$

8) $\dfrac{8}{5} = \dfrac{8 \times 2}{5 \times 2} = \dfrac{16}{10} = 1.6$

9) $\dfrac{8}{125} = \dfrac{8 \times 8}{125 \times 8} = \dfrac{64}{1000} = 0.064$

10) $\dfrac{301}{250} = \dfrac{301 \times 4}{250 \times 4} = \dfrac{1204}{1000} = 1.204$

11) $\dfrac{7}{10} = 0.7$

12) $\dfrac{41}{200} = \dfrac{41 \times 5}{200 \times 5} = \dfrac{205}{1000} = 0.205$

4.6 Repeating Decimals

1) $\dfrac{2}{3} = \dfrac{2 \times 3}{3 \times 3} = \dfrac{6}{9} = 0.\overline{6}$

2) $\dfrac{8}{11} = \dfrac{8 \times 9}{11 \times 9} = \dfrac{72}{99} = 0.\overline{72}$

3) $\frac{16}{333} = \frac{16 \times 3}{333 \times 3} = \frac{48}{999} = 0.\overline{048}$ (the zero after the decimal point is repeating)

4) $\frac{5}{9} = 0.\overline{5}$

5) $\frac{1}{90} = \frac{1}{10} \times \frac{1}{9} = \frac{1}{10} \times 0.\overline{1} = 0.0\overline{1}$ (only the 1 is repeating)

6) $\frac{5}{3} = 1\frac{2}{3} = 1 + \frac{2}{3} = 1 + \frac{2 \times 3}{3 \times 3} = 1 + \frac{6}{9} = 1 + 0.\overline{6} = 1.\overline{6}$

7) $\frac{43}{99} = 0.\overline{43}$

8) $\frac{31}{111} = \frac{31 \times 9}{111 \times 9} = \frac{279}{999} = 0.\overline{279}$

9) $\frac{7}{30} = \frac{1}{10} \times \frac{7}{3} = \frac{1}{10} \times 2\frac{1}{3} = \frac{1}{10} \times \left(2 + \frac{1}{3}\right) = \frac{1}{10} \times \left(2 + \frac{1 \times 3}{3 \times 3}\right) = \frac{1}{10} \times \left(2 + \frac{3}{9}\right) = \frac{1}{10} \times (2 + 0.\overline{3}) =$

$\frac{1}{10} \times 2.\overline{3} = 0.2\overline{3}$ (the 3 repeats, but the 2 does not)

10) $\frac{14}{9} = 1\frac{5}{9} = 1 + \frac{5}{9} = 1 + 0.\overline{5} = 1.\overline{5}$

11) $\frac{2}{45} = \frac{2 \times 2}{45 \times 2} = \frac{4}{90} = \frac{1}{10} \times \frac{4}{9} = \frac{1}{10} \times 0.\overline{4} = 0.0\overline{4}$ (only the 4 is repeating)

12) $\frac{4}{15} = \frac{4 \times 6}{15 \times 6} = \frac{24}{90} = \frac{1}{10} \times \frac{24}{9} = \frac{1}{10} \times 2\frac{6}{9} = \frac{1}{10} \times \left(2 + \frac{6}{9}\right) = \frac{1}{10} \times (2 + 0.\overline{6}) = \frac{1}{10} \times 2.\overline{6} = 0.2\overline{6}$ (only

the 6 repeats; the 2 doesn't)

13) $\frac{28}{11} = 2\frac{6}{11} = 2 + \frac{6}{11} = 2 + \frac{6 \times 9}{11 \times 9} = 2 + \frac{54}{99} = 2 + 0.\overline{54} = 2.\overline{54}$

14) $\frac{1}{900} = \frac{1}{100} \times \frac{1}{9} = \frac{1}{100} \times 0.\overline{1} = 0.00\overline{1}$ (only the 1 repeats indefinitely)

15) $\frac{5}{6} = \frac{5 \times 15}{6 \times 15} = \frac{75}{90} = \frac{1}{10} \times \frac{75}{9} = \frac{1}{10} \times 8\frac{3}{9} = \frac{1}{10} \times \left(8 + \frac{3}{9}\right) = \frac{1}{10} \times (8 + 0.\overline{3}) = \frac{1}{10} \times 8.\overline{3} = 0.8\overline{3}$ (only

the 3 repeats; the 8 doesn't)

16) $\frac{19}{180} = \frac{1}{10} \times \frac{19}{18} = \frac{1}{10} \times 1\frac{1}{18} = \frac{1}{10} \times \left(1 + \frac{1}{18}\right) = \frac{1}{10} \times \left(1 + \frac{1 \times 5}{18 \times 5}\right) = \frac{1}{10} \times \left(1 + \frac{5}{90}\right) = \frac{1}{10} \times \left(1 + \frac{1}{10} \times \frac{5}{9}\right)$

$= \frac{1}{10} \times \left(1 + \frac{1}{10} \times 0.\overline{5}\right) = \frac{1}{10} \times (1 + 0.0\overline{5}) = \frac{1}{10} \times 1.0\overline{5} = 0.10\overline{5}$ (only the 5 repeats)

17) $\frac{301}{300} = 1\frac{1}{300} = 1 + \frac{1}{300} = 1 + \frac{1}{100} \times \frac{1}{3} = 1 + \frac{1}{100} \times \frac{1 \times 3}{3 \times 3} = 1 + \frac{1}{100} \times \frac{3}{9} = 1 + \frac{1}{100} \times 0.\overline{3} = 1 +$

$0.00\overline{3} = 1.00\overline{3}$ (only the 3 repeats indefinitely)

18) $\frac{11}{12} = \frac{11 \times 75}{12 \times 75} = \frac{825}{900} = \frac{1}{100} \times \frac{825}{9} = \frac{1}{100} \times 91\frac{6}{9} = \frac{1}{100} \times \left(91 + \frac{6}{9}\right) = \frac{1}{100} \times (91 + 0.\overline{6}) = \frac{1}{100} \times 91.\overline{6}$

$= 0.91\overline{6}$ (only the 6 repeats)

19) $\frac{7}{60} = \frac{7 \times 15}{60 \times 15} = \frac{105}{900} = \frac{1}{100} \times \frac{105}{9} = \frac{1}{100} \times 11\frac{6}{9} = \frac{1}{100} \times \left(11 + \frac{6}{9}\right) = \frac{1}{100} \times (11 + 0.\overline{6}) = \frac{1}{100} \times 11.\overline{6}$

$= 0.11\overline{6}$ (only the 6 repeats indefinitely)

20) $\frac{8}{27} = \frac{8 \times 37}{27 \times 37} = \frac{296}{999} = 0.\overline{296}$

4.7 Adding and Subtracting Decimals

1)
$$\begin{array}{r} 4.8 \\ +6.5 \\ \hline 11.3 \end{array}$$

2)
$$\begin{array}{r} 11.4 \\ -3.8 \\ \hline 7.6 \end{array}$$

3)
$$\begin{array}{r} 0.80 \\ -0.54 \\ \hline 0.26 \end{array}$$

4)
$$\begin{array}{r} 2.74 \\ +4.36 \\ \hline 7.1 \end{array}$$

5)
$$\begin{array}{r} 0.35 \\ +0.47 \\ \hline 0.82 \end{array}$$

6)
$$\begin{array}{r} 0.072 \\ -0.050 \\ \hline 0.022 \end{array}$$

7)
$$\begin{array}{r} 4.20 \\ -1.34 \\ \hline 2.86 \end{array}$$

8)
$$\begin{array}{r} 4.580 \\ +0.892 \\ \hline 5.472 \end{array}$$

9)
$$\begin{array}{r} 0.0024 \\ +0.0073 \\ \hline 0.0097 \end{array}$$

10)
$$\begin{array}{r} 4.00 \\ -1.83 \\ \hline 2.17 \end{array}$$

11)
$$\begin{array}{r} 21.4 \\ -7.9 \\ \hline 13.5 \end{array}$$

12)
$$\begin{array}{r} 0.164 \\ +0.380 \\ \hline 0.544 \end{array}$$

4.8 Multiplying Decimals

1) $0.4 \times 0.8 = 0.32$ (since $4 \times 8 = 32$ and there are 2 decimal places)

2) $0.03 \times 0.2 = 0.006$ (since $2 \times 3 = 6$ and there are 3 decimal places)

3) $0.6 \times 0.25 = 0.15$ (since $6 \times 25 = 150$ and there are 3 decimal places before dropping the trailing zero; note that $0.150 = 0.15$)

4) $0.11 \times 0.12 = 0.0132$ (since $11 \times 12 = 132$ and there are 4 decimal places)

5) $1.5 \times 2.4 = 3.6$ (since $15 \times 24 = 360$ and there are 2 decimal places before dropping the trailing zero; note that $3.60 = 3.6$)

6) $7 \times 3.5 = 24.5$ (since $7 \times 35 = 245$ and there is 1 decimal place)

7) $0.06 \times 0.05 = 0.003$ (since $6 \times 5 = 30$ and there are 4 decimal places before dropping the trailing zero; note that $0.0030 = 0.003$)

8) $0.007 \times 0.09 = 0.00063$ (since $7 \times 9 = 63$ and there are 5 decimal places)

9) $0.004 \times 0.003 = 0.000012$ (since $4 \times 3 = 12$ and there are 6 decimal places)

10) $1.6 \times 0.06 = 0.096$ (since $16 \times 6 = 96$ and there are 3 decimal places)

11) $14.5 \times 0.08 = 1.16$ (since $145 \times 8 = 1160$ and there are 3 decimal places before dropping the trailing zero; note that $1.160 = 1.16$)

12) $0.05 \times 0.004 = 0.0002$ (since $5 \times 4 = 20$ and there are 5 decimal places before dropping the trailing zero; note that $0.00020 = 0.0002$)

4.9 Dividing Decimals

1) $0.48 \div 0.8 = \frac{0.48}{0.8} = \frac{0.48 \times 100}{0.8 \times 100} = \frac{48}{80} = \frac{48 \div 8}{80 \div 8} = \frac{6}{10} = 0.6$

Check the answer: $0.6 \times 0.8 = 0.48$

2) $1.7 \div 5 = \frac{1.7}{5} = \frac{1.7 \times 10}{5 \times 10} = \frac{17}{50} = \frac{17 \times 2}{50 \times 2} = \frac{34}{100} = 0.34$

Check the answer: $0.34 \times 5 = 1.7$

3) $0.032 \div 0.05 = \frac{0.032}{0.05} = \frac{0.032 \times 1000}{0.05 \times 1000} = \frac{32}{50} = \frac{32 \times 2}{50 \times 2} = \frac{64}{100} = 0.64$

Check the answer: $0.64 \times 0.05 = 0.032$

4) $0.6 \div 0.8 = \frac{0.6}{0.8} = \frac{0.6 \times 10}{0.8 \times 10} = \frac{6}{8} = \frac{6 \div 2}{8 \div 2} = \frac{3}{4} = \frac{3 \times 25}{4 \times 25} = \frac{75}{100} = 0.75$

Check the answer: $0.75 \times 0.8 = 0.6$

5) $4.2 \div 0.6 = \frac{4.2}{0.6} = \frac{4.2 \times 10}{0.6 \times 10} = \frac{42}{6} = \frac{42 \div 6}{6 \div 6} = \frac{7}{1} = 7$

Check the answer: $7 \times 0.6 = 4.2$

6) $0.0016 \div 0.0032 = \frac{0.0016}{0.0032} = \frac{0.0016 \times 10,000}{0.0032 \times 10,000} = \frac{16}{32} = \frac{16 \div 16}{32 \div 16} = \frac{1}{2} = \frac{1 \times 5}{2 \times 5} = \frac{5}{10} = 0.5$

Check the answer: $0.5 \times 0.0032 = 0.0016$

7) $0.08 \div 0.4 = \frac{0.08}{0.4} = \frac{0.08 \times 100}{0.4 \times 100} = \frac{8}{40} = \frac{8 \div 4}{40 \div 4} = \frac{2}{10} = 0.2$

Check the answer: $0.2 \times 0.4 = 0.08$

8) $1.2 \div 9.6 = \frac{1.2}{9.6} = \frac{1.2 \times 10}{9.6 \times 10} = \frac{12}{96} = \frac{12 \div 12}{96 \div 12} = \frac{1}{8} = \frac{1 \times 125}{8 \times 125} = \frac{125}{1000} = 0.125$

Check the answer: $0.125 \times 9.6 = 1.2$

4.10 Powers of Decimals

1) $0.9^2 = 0.9 \times 0.9 = 0.81$

2) $0.3^3 = 0.3 \times 0.3 \times 0.3 = 0.09 \times 0.3 = 0.027$

3) $\sqrt{0.36} = \pm 0.6$ since $0.6^2 = 0.6 \times 0.6 = 0.36$

4) $\sqrt[3]{0.027} = 0.3$ since $0.3^3 = 0.3 \times 0.3 \times 0.3 = 0.09 \times 0.3 = 0.027$

5) $0.5^{-1} = \frac{1}{0.5} = \frac{1 \times 2}{0.5 \times 2} = \frac{2}{1} = 2$

6) $2.5^{-1} = \frac{1}{2.5} = \frac{1 \times 4}{2.5 \times 4} = \frac{4}{10} = 0.4$

7) $0.001^{1/3} = \left(\sqrt[3]{0.001} \right)^1 = 0.1$

8) $0.16^{-1/2} = \left(\sqrt[2]{0.16} \right)^{-1} = (\pm 0.4)^{-1} = \frac{1}{\pm 0.4} = \pm 2.5$

9) $0.04^{3/2} = \left(\sqrt{0.04}\right)^3 = (\pm 0.2)^3 = \pm 0.2 \times 0.2 \times 0.2 = \pm 0.04 \times 0.2 = \pm 0.008$

10) $0.0001^{-3/4} = \left(\sqrt[4]{0.0001}\right)^{-3} = (\pm 0.1)^{-3} = \frac{1}{(\pm 0.1)^3} = \frac{1}{\pm 0.1 \times 0.1 \times 0.1} = \frac{1}{\pm 0.01 \times 0.1} = \frac{1}{\pm 0.001} = \pm 1000$

4.11 Dollars and Cents

1) $\$3.25 \div \$0.05 = \frac{3.25}{0.05} = \frac{3.25 \times 100}{.05 \times 100} = \frac{325}{5} = 325 \div 5 = 65$

2) $\$7.60 \div \$0.10 = \frac{7.60}{0.10} = \frac{7.60 \times 10}{.10 \times 10} = \frac{76}{1} = 76$

3) $\$1.63 \div \$0.01 = \frac{1.63}{0.01} = \frac{1.63 \times 100}{.01 \times 100} = \frac{163}{1} = 163 \div 1 = 163$

4) $\$8.75 \div \$0.25 = \frac{8.75}{0.25} = \frac{8.75 \times 100}{.25 \times 100} = \frac{875}{25} = 875 \div 25 = 35$

5) $\$6.00 \div \$0.05 = \frac{6.00}{0.05} = \frac{6.00 \times 100}{.05 \times 100} = \frac{600}{5} = 600 \div 5 = 120$

6) $15 \times \$0.25 + 7 \times \$0.10 + 5 \times \$0.05 + 18 \times \$0.01 = \$3.75 + \$0.70 + \$0.25 + \$0.18 = \$4.88$

7) $7 \times \$0.8 + 2 \times \$2.25 = \$5.6 + \$4.5 = \$10.10$ (which is equivalent to $10.1)

8) $5 \times \$0.18 + 3 \times \$0.42 + 2 \times \$1.25 = \$0.9 + \$1.26 + \$2.5 = \$4.66$

9) $12 \times \$12.5 + 5 \times \$16.25 + 8 \times \$5.75 = \$150 + \$81.25 + 46 = \$277.25$

5 Percents

5.1 Converting Decimals to Percents

1) $0.91 = 0.91 \times 100\% = 91\%$

2) $0.03 = 0.03 \times 100\% = 3\%$

3) $2.2 = 2.2 \times 100\% = 220\%$

4) $0.362 = 0.362 \times 100\% = 36.2\%$

5) $0.5 = 0.5 \times 100\% = 50\%$

6) $0.005 = 0.005 \times 100\% = 0.5\%$

7) $1.23 = 1.23 \times 100\% = 123\%$

8) $0.048 = 0.048 \times 100\% = 4.8\%$

9) $0.125 = 0.125 \times 100\% = 12.5\%$

10) $0.2468 = 0.2468 \times 100\% = 24.68\%$

11) $3 = 3 \times 100\% = 300\%$

12) $0.0841 = 0.0841 \times 100\% = 8.41\%$

13) $0.101 = 0.101 \times 100\% = 10.1\%$

14) $0.74 = 0.74 \times 100\% = 74\%$

15) $0.9 = 0.9 \times 100\% = 90\%$

16) $0.025 = 0.025 \times 100\% = 2.5\%$

17) $1.414 = 1.414 \times 100\% = 141.4\%$

18) $0.0008 = 0.0008 \times 100\% = 0.08\%$

5.2 Converting Percents to Decimals

1) $82\% = \frac{82\%}{100\%} = 0.82$

2) $6\% = \frac{6\%}{100\%} = 0.06$

3) $0.7\% = \frac{0.7\%}{100\%} = 0.007$

4) $135\% = \frac{135\%}{100\%} = 1.35$

5) $10\% = \frac{10\%}{100\%} = 0.1$

6) $100\% = \frac{100\%}{100\%} = 1$

7) $1\% = \frac{1\%}{100\%} = 0.01$

8) $0.1\% = \frac{0.1\%}{100\%} = 0.001$

9) $0.04\% = \frac{0.04\%}{100\%} = 0.0004$

10) $32.6\% = \frac{32.6\%}{100\%} = 0.326$

11) $49\% = \frac{49\%}{100\%} = 0.49$

12) $700\% = \frac{700\%}{100\%} = 7$

13) $0.006\% = \frac{0.006\%}{100\%} = 0.00006$

14) $1.357\% = \frac{1.357\%}{100\%} = 0.01357$

15) $0.738\% = \frac{0.738\%}{100\%} = 0.00738$

16) $250\% = \frac{250\%}{100\%} = 2.5$

17) $4.2\% = \frac{4.2\%}{100\%} = 0.042$

18) $0.0375\% = \frac{0.0375\%}{100\%} = 0.000375$

5.3 Converting Percents to Fractions

1) $50\% = \frac{50}{100} = \frac{50 \div 50}{100 \div 50} = \frac{1}{2}$

2) $8\% = \frac{8}{100} = \frac{8 \div 4}{100 \div 4} = \frac{2}{25}$

3) $80\% = \frac{80}{100} = \frac{80 \div 20}{100 \div 20} = \frac{4}{5}$

4) $150\% = \frac{150}{100} = \frac{150 \div 50}{100 \div 50} = \frac{3}{2}$ or $1\frac{1}{2}$

5) $10\% = \frac{10}{100} = \frac{10 \div 10}{100 \div 10} = \frac{1}{10}$

6) $75\% = \frac{75}{100} = \frac{75 \div 25}{100 \div 25} = \frac{3}{4}$

7) $0.5\% = \frac{0.5}{100} = \frac{0.5 \times 2}{100 \times 2} = \frac{1}{200}$

8) $36\% = \frac{36}{100} = \frac{36 \div 4}{100 \div 4} = \frac{9}{25}$

9) $125\% = \frac{125}{100} = \frac{125 \div 25}{100 \div 25} = \frac{5}{4}$ or $1\frac{1}{4}$

10) $4\% = \frac{4}{100} = \frac{4 \div 4}{100 \div 4} = \frac{1}{25}$

11) $12.5\% = \frac{12.5}{100} = \frac{12.5 \div 12.5}{100 \div 12.5} = \frac{1}{8}$

12) $0.2\% = \frac{0.2}{100} = \frac{0.2 \times 5}{100 \times 5} = \frac{1}{500}$

5.4 Converting Fractions to Percents

1) $\frac{1}{4} = \frac{1 \times 25}{4 \times 25} = \frac{25}{100} = 25\%$

2) $\frac{9}{20} = \frac{9 \times 5}{20 \times 5} = \frac{45}{100} = 45\%$

3) $\frac{2}{5} = \frac{2 \times 20}{5 \times 20} = \frac{40}{100} = 40\%$

4) $\frac{5}{2} = \frac{5 \times 50}{2 \times 50} = \frac{250}{100} = 250\%$

5) $\frac{11}{20} = \frac{11 \times 5}{20 \times 5} = \frac{55}{100} = 55\%$

6) $\frac{9}{10} = \frac{9 \times 10}{10 \times 10} = \frac{90}{100} = 90\%$

7) $\frac{12}{25} = \frac{12 \times 4}{25 \times 4} = \frac{48}{100} = 48\%$

8) $\frac{7}{5} = \frac{7 \times 20}{5 \times 20} = \frac{140}{100} = 140\%$

9) $\frac{9}{50} = \frac{9 \times 2}{50 \times 2} = \frac{18}{100} = 18\%$

10) $\frac{1}{8} = \frac{1 \times 12.5}{8 \times 12.5} = \frac{12.5}{100} = 12.5\%$

5.5 Discounts and Sales Tax

1) discount rate: $\frac{15\%}{100\%} = 0.15$

amount of discount: $0.15 \times \$80 = \$12$

total cost: $\$80 - \$12 = \$68$

2) tax rate: $\frac{9\%}{100\%} = 0.09$

amount of tax: $0.09 \times \$5 = \$0.45$

total cost: $\$5 + \$0.45 = \$5.45$

3) discount rate: $\frac{25\%}{100\%} = 0.25$, tax rate: $\frac{10\%}{100\%} = 0.1$

amount of discount: $0.25 \times \$24 = \$6$

sale price: $\$24 - \$6 = \$18$

amount of tax: $0.1 \times \$18 = \$1.8$

total cost: $\$18 + \$1.8 = \$19.8$ (it's the same as $19.80)

5.6 Simple Interest

1) interest rate: $\frac{6\%}{100\%} = 0.06$

amount of interest: $0.06 \times \$3000 = \$180$

new balance: $\$3000 + \$180 = \$3180$

2) interest rate: $\frac{7\%}{100\%} = 0.07$

amount of interest: $0.07 \times \$1500 = \$105$

total amount paid: $\$1500 + \$105 = \$1605$

5.7 Percent Increase and Decrease

1) rate of increase: $\frac{15\%}{100\%} = 0.15$

factor: $1 + 0.15 = 1.15$

new value: $1.15 \times \$42 = \$48.3$ (it's the same as $48.30)

(Alternate solution: $0.15 \times \$42 = \$6.3$ and $\$42 + \$6.3 = \$48.3$)

2) rate of decrease: $\frac{8\%}{100\%} = 0.08$

factor: $1 - 0.08 = 0.92$

new value: $0.92 \times 3600 = 3312$

(Alternate solution: $0.08 \times 3600 = 288$ and $3600 - 288 = 3312$)

3) rate of increase: $\frac{10\%}{100\%} = 0.1$

factor: $1 + 0.1 = 1.1$

new value: $1.1 \times 80\% = 88\%$

(Alternate solution: $0.1 \times 80\% = 8\%$ and $80\% + 8\% = 88\%$)

4) rates of increase/decrease: $\frac{9\%}{100\%} = 0.09$ and $\frac{4\%}{100\%} = 0.04$

factors: $1 + 0.09 = 1.09$ and $1 - 0.04 = 0.96$

new values: $1.09 \times \$600 = \$654$ and $0.96 \times \$600 = \$576$

difference: $\$654 - \$576 = \$78$

(Alternate solution: $0.09 \times \$600 = \$54$, $-0.04 \times \$600 = -\$24$, and $\$54 - (-\$24) = \$54 + \$24 = \$78$)

6 Working with Expressions

6.1 Operations with Variables

1) $7 - x$ means 7 minus x (or subtract x from 7)

2) $\frac{x}{5}$ means x divided by 5 (or one-fifth of x)

3) $4x + 1$ means 4 times x plus 1 (or multiply 4 by x and then add 1)

4) $0.75x$ means 75% of x (or 0.75 times x or three-quarters of x)

5) $\frac{2}{x}$ means 2 divided by x

6) $x + yz$ means x plus y times z (or add x to the product of y and z)

7) $\sqrt{3x}$ means first multiply x by 3 and then take the square root

6.2 Combine Like Terms

1) $5x + 8 - 3x + 6 = (5x - 3x) + (8 + 6) = 2x + 14$

2) $12x + 6 + 13x + 3 = (12x + 13x) + (6 + 3) = 25x + 9$

3) $x^2 + 8 + x^2 + 9 = (1x^2 + 1x^2) + (8 + 9) = 2x^2 + 17$

4) $15x - 8 + 10x - 6 + 8x - 4 = (15x + 10x + 8x) + (-8 - 6 - 4) = 33x - 18$

5) $3x + x + 11x = (3x + 1x + 11x) = 15x$

6) $8x^2 + 10 - 4x^2 - 7 = (8x^2 - 4x^2) + (10 - 7) = 4x^2 + 3$

7) $12 + 14x - 2x - 6 - 2x - 5 = (14x - 2x - 2x) + (12 - 6 - 5) = 10x + 1$

8) $x - 3 + x - 1 + 2x - 3 = (1x + 1x + 2x) + (-3 - 1 - 3) = 4x - 7$

9) $16x^2 + 14 - 3x^2 - 12 - 2x^2 + 13 = (16x^2 - 3x^2 - 2x^2) + (14 - 12 + 13) = 11x^2 + 15$

10) $3x^2 + 14 + 4x^2 + 13 + 2x^2 - 11 = (3x^2 + 4x^2 + 2x^2) + (14 + 13 - 11) = 9x^2 + 16$

11) $8 - 17 + 4x^2 + 2x^2 + 11 + 11 = (4x^2 + 2x^2) + (8 - 17 + 11 + 11) = 6x^2 + 13$

12) $x - 14 - x + 8x + 12 - 1 = (1x - 1x + 8x) + (-14 + 12 - 1) = 8x - 3$

13) $6x^3 + 2x^3 + x + x + 22 - 13 = (6x^3 + 2x^3) + (1x + 1x) + (22 - 13) = 8x^3 + 2x + 9$

14) $7x + 4x + 6x - 12 + 4 = (7x + 4x + 6x) + (-12 + 4) = 17x - 8$

15) $4x^2 + 9 + 6x^2 + 19x + 19 = (4x^2 + 6x^2) + 19x + (9 + 19) = 10x^2 + 19x + 28$

16) $x^3 + 4x + 7 + 6x^3 + 9x + 12 = (1x^3 + 6x^3) + (4x + 9x) + (7 + 12) = 7x^3 + 13x + 19$

6.3 Exploring Expressions

1) $(3)^2 - 2(3) + 5 = 9 - 6 + 5 = 3 + 5 = 8$

2) $\frac{19-4}{3} = \frac{15}{3} = 5$

3) $7(6) - 3(6) = 42 - 18 = 24$ agrees with $4(6) = 24$

4) $3(10)^2 + 2(10)^2 = 300 + 200 = 500$ agrees with $5(10)^2 = 500$

5) $3(7) - 7 = 21 - 7 = 14$ agrees with $2(7) = 14$

6) $3(-2) - 4 = -6 - 4 = -10$

7) $(-3)^3 + 2(-3)^2 - 4(-3) - 7 = -27 + 2(9) + 12 - 7 = -27 + 18 + 12 - 7 = -34 + 30 = -4$

8) $8(-5) - (-5) + 2(-5) = -40 + 5 - 10 = -50 + 5 = -45$ agrees with $9(-5) = -45$

9) $3(8) - 2(3) = 24 - 6 = 18$

10) $(2)^3(3)^2 = (8)(9) = 72$

11) $\frac{9+6}{7-4} = \frac{15}{3} = 5$

12) $8\left(\frac{1}{2}\right) - 2\left(\frac{1}{2}\right) = 4 - 1 = 3$ agrees with $6\left(\frac{1}{2}\right) = 3$

13) $\left(\frac{3}{2}\right)^2 + 2\left(\frac{3}{2}\right)^2 = \frac{9}{4} + 2\left(\frac{9}{4}\right) = \frac{9}{4} + \frac{18}{4} = \frac{27}{4}$ or $6\frac{3}{4}$ agrees with $3\left(\frac{3}{2}\right)^2 = \frac{27}{4}$ or $6\frac{3}{4}$

6.4 Exploring Powers of Variables

1) $3^2 \cdot 3^4 = (3 \cdot 3)(3 \cdot 3 \cdot 3 \cdot 3) = (9)(81) = 729$ agrees with $3^{2+4} = 3^6 = 729$

2) $10^3 \cdot 10^4 = (10 \cdot 10 \cdot 10)(10 \cdot 10 \cdot 10 \cdot 10) = (1000)(10,000) = 10,000,000$ agrees with 10^{3+4} $= 10^7 = 10,000,000$

3) $4^{-2} \cdot 4^3 = \frac{1}{16} \cdot 64 = 4$ agrees with $4^{-2+3} = 4^1 = 4$

4) $10^{-3} \cdot 10^3 = \frac{1}{10 \cdot 10 \cdot 10}(10 \cdot 10 \cdot 10) = \frac{1000}{1000} = 1$ agrees with $10^{-3+3} = 10^0 = 1$

5) $\frac{2^5}{2^3} = \frac{2 \cdot 2 \cdot 2 \cdot 2 \cdot 2}{2 \cdot 2 \cdot 2} = \frac{32}{8} = 4$ agrees with $2^{5-3} = 2^2 = 4$

6) $\frac{3^6}{3^{-2}} = \frac{3 \cdot 3 \cdot 3 \cdot 3 \cdot 3 \cdot 3}{\frac{1}{3^2}} = \frac{729}{\frac{1}{9}} = 729 \div \frac{1}{9} = 729 \times \frac{9}{1} = \frac{729 \times 9}{1} = \frac{6561}{1} = 6561$ agrees with $3^{6-(-2)} =$
$3^{6+2} = 3^8 = 6561$

7) $(2^3)^4 = (8)^4 = 4096$ agrees with $2^{12} = 4096$

6.5 Working with Powers of Variables

1) $x^3x^2 = x^{3+2} = x^5$

2) $x^8x^6 = x^{8+6} = x^{14}$

3) $x^4x = x^4x^1 = x^{4+1} = x^5$

4) $x^4x^2x^0 = x^{4+2+0} = x^6$

5) $x^5x^{-3} = x^{5+(-3)} = x^{5-3} = x^2$

6) $x^{-7}x^8 = x^{-7+8} = x^1 = x$

7) $x^{-9}x^{-8} = x^{-9+(-8)} = x^{-17} = \dfrac{1}{x^{17}}$

8) $x^3x^{-4}x = x^{3+(-4)+1} = x^0 = 1$

9) $\dfrac{x^7}{x^3} = x^{7-3} = x^4$

10) $\dfrac{x^{12}}{x^5} = x^{12-5} = x^7$

11) $\dfrac{x^4}{x^4} = x^{4-4} = x^0 = 1$

12) $\dfrac{x^6}{x^{-6}} = x^{6-(-6)} = x^{6+6} = x^{12}$

13) $\dfrac{x^{-2}}{x^3} = x^{-2-3} = x^{-5} = \dfrac{1}{x^5}$

14) $\dfrac{x}{x^2} = \dfrac{x^1}{x^2} = x^{1-2} = x^{-1} = \dfrac{1}{x}$

15) $\dfrac{x^{-7}}{x^{-4}} = x^{-7-(-4)} = x^{-7+4} = x^{-3} = \dfrac{1}{x^3}$

16) $\dfrac{x^{-5}}{x^{-9}} = x^{-5-(-9)} = x^{-5+9} = x^4$

17) $x^{-3} = \dfrac{1}{x^3}$

18) $\dfrac{1}{x^{-3}} = \dfrac{x^0}{x^{-3}} = x^{0-(-3)} = x^{0+3} = x^3$

19) $\dfrac{x^2}{1} = x^2$

20) $\dfrac{x^{-4}}{x^0} = x^{-4-0} = x^{-4} = \dfrac{1}{x^4}$

21) $(x^2)^4 = x^{2\cdot4} = x^8$

22) $(x^5)^6 = x^{5\cdot6} = x^{30}$

23) $(x^{-2})^3 = x^{-2\cdot3} = x^{-6} = \dfrac{1}{x^6}$

24) $(x^{-3})^{-4} = x^{(-3)(-4)} = x^{3\cdot4} = x^{12}$

25) $(x^{-1})^{-1} = x^{(-1)(-1)} = x^1 = x$

26) $x^{-1}x^{-1} = x^{-1+(-1)} = x^{-2} = \dfrac{1}{x^2}$

27) $\dfrac{x^4x^8}{x^2x^3} = \dfrac{x^{4+8}}{x^{2+3}} = \dfrac{x^{12}}{x^5} = x^{12-5} = x^7$

28) $\dfrac{x^6x^{-3}}{x^5x^{-4}} = \dfrac{x^{6+(-3)}}{x^{5+(-4)}} = \dfrac{x^3}{x^1} = x^{3-1} = x^2$

29) $\dfrac{x^2x}{x^{-3}} = \dfrac{x^{2+1}}{x^{-3}} = \dfrac{x^3}{x^{-3}} = x^{3-(-3)} = x^{3+3} = x^6$

30) $\dfrac{x^3x^{-7}}{x^5} = \dfrac{x^{3+(-7)}}{x^5} = \dfrac{x^{-4}}{x^5} = x^{-4-5} = x^{-9} = \dfrac{1}{x^9}$

31) $\dfrac{x^8x^{-8}}{x^{-1}} = \dfrac{x^{8+(-8)}}{x^{-1}} = \dfrac{x^0}{x^{-1}} = x^{0-(-1)} = x^{0+1} = x$

32) $\left(\dfrac{x^5}{x^2}\right)^3 = (x^{5-2})^3 = (x^3)^3 = x^{3\cdot3} = x^9$

33) $\left(\dfrac{x^{-2}}{x^{-3}}\right)^4 = \left[x^{-2-(-3)}\right]^4 = (x^1)^4 = x^{1\cdot4} = x^4$

34) $\left(\dfrac{x^9x^6}{x^8}\right)^2 = \left(\dfrac{x^{9+6}}{x^8}\right)^2 = \left(\dfrac{x^{15}}{x^8}\right)^2 = (x^7)^2 = x^{14}$

35) $\left(\dfrac{x^4}{x^5}\right)^{-1} = (x^{4-5})^{-1} = (x^{-1})^{-1} = x^{(-1)(-1)} = x^1 = x$

36) $\left(\dfrac{x^5}{x}\right)^{-2} = \left(\dfrac{x^5}{x^1}\right)^{-2} = (x^{5-1})^{-2} = (x^4)^{-2} = x^{4(-2)} = x^{-8} = \dfrac{1}{x^8}$

37) $\left(\dfrac{1}{x^3}\right)^{-4} = \left(\dfrac{x^0}{x^3}\right)^{-4} = (x^{0-3})^{-4} = (x^{-3})^{-4} = x^{-3(-4)} = x^{12}$

38) $\left(\dfrac{1}{x^{-4}}\right)^{-3} = \left(\dfrac{x^0}{x^{-4}}\right)^{-3} = \left[x^{0-(-4)}\right]^{-3} = (x^4)^{-3} = x^{4(-3)} = x^{-12} = \dfrac{1}{x^{12}}$

39) $\left(\dfrac{x^3}{x^{-2}}\right)^{-5} = \left[x^{3-(-2)}\right]^{-5} = (x^{3+2})^{-5} = (x^5)^{-5} = x^{5(-5)} = x^{-25} = \dfrac{1}{x^{25}}$

40) $\left(\frac{x^{-7}}{x^{-3}}\right)^{-1} = \left[x^{-7-(-3)}\right]^{-1} = (x^{-7+3})^{-1} = (x^{-4})^{-1} = x^{-4(-1)} = x^4$

41) $\left(\frac{x^{-3}}{x}\right)^{-2} = \left(\frac{x^{-3}}{x^1}\right)^{-2} = (x^{-3-1})^{-2} = (x^{-4})^{-2} = x^{-4(-2)} = x^8$

42) $\left(\frac{x^{-2}}{x^{-5}}\right)^{-4} = \left[x^{-2-(-5)}\right]^{-4} = (x^{-2+5})^{-4} = (x^3)^{-4} = x^{3(-4)} = x^{-12} = \frac{1}{x^{12}}$

43) $\left(\frac{x^8 x^7}{x^5 x^4}\right)^3 = \left(\frac{x^{8+7}}{x^{5+4}}\right)^3 = \left(\frac{x^{15}}{x^9}\right)^3 = (x^{15-9})^3 = (x^6)^3 = x^{6\cdot3} = x^{18}$

44) $\left(\frac{x^5 x^{-3}}{x^2 x^{-4}}\right)^{-2} = \left(\frac{x^{5+(-3)}}{x^{2+(-4)}}\right)^{-2} = \left(\frac{x^2}{x^{-2}}\right)^{-2} = \left[x^{2-(-2)}\right]^{-2} = (x^4)^{-2} = x^{4(-2)} = x^{-8} = \frac{1}{x^8}$

45) $(5x)^3 = 5^3 x^{1\cdot3} = 125x^3$

46) $(2x^3)^6 = 2^6 x^{3\cdot6} = 64x^{18}$

47) $(-x^2)^5 = (-1)^5 x^{2\cdot5} = (-1)x^{10} = -x^{10}$

48) $(-x^7)^8 = (-1)^8 x^{7\cdot8} = (1)x^{56} = x^{56}$

49) $(8x^9)^{-2} = 8^{-2} x^{9(-2)} = 8^{-2} x^{-18} = \frac{1}{8^2}\frac{1}{x^{18}} = \frac{1}{64x^{18}}$

50) $(5x^0)^4 = 5^4 x^{0\cdot4} = 5^4 x^0 = 5^4(1) = 5^4 = 625$

51) $(-2x^{-3})^6 = (-2)^6 x^{-3\cdot6} = 64x^{-18} = \frac{64}{x^{18}}$

52) $(3x^2)^{-2} = 3^{-2} x^{2(-2)} = 3^{-2} x^{-4} = \frac{1}{3^2}\frac{1}{x^4} = \frac{1}{9x^4}$

53) $(7x^6)^0 = 7^0 x^{6\cdot0} = 7^0 x^0 = (1)(1) = 1$

54) $(-4x^{-5})^{-3} = (-4)^{-3} x^{-5(-3)} = \frac{1}{(-4)^3}x^{15} = -\frac{x^{15}}{64}$

55) $(x^7 y^6)^8 = x^{7\cdot8} y^{6\cdot8} = x^{56} y^{48}$

56) $(9xy)^2 = 9^2 x^2 y^2 = 81x^2 y^2$

57) $(2x^4 y^6)^5 = 2^5 x^{4\cdot5} y^{6\cdot5} = 32x^{20} y^{30}$

58) $(-x^6 y^3)^3 = (-1)^3 x^{6\cdot3} y^{3\cdot3} = -x^{18} y^9$

59) $(6x^8 y^8)^2 = 6^2 x^{8\cdot2} y^{8\cdot2} = 36x^{16} y^{16}$

60) $(8x^5 y^7)^1 = 8^1 x^{5\cdot1} y^{7\cdot1} = 8x^5 y^7$

61) $(-2x^{-7} y^{-2})^6 = (-2)^6 x^{-7\cdot6} y^{-2\cdot6} = 64x^{-42} y^{-12} = \frac{64}{x^{42} y^{12}}$

62) $(3x^4 y^{-1})^4 = 3^4 x^{4\cdot4} y^{-1\cdot4} = 81x^{16} y^{-4} = \frac{81x^{16}}{y^4}$

63) $(x^3 y^{-2})^{-1} = x^{3(-1)} y^{(-2)(-1)} = x^{-3} y^2 = \frac{y^2}{x^3}$

64) $(-x^5 y^{-5})^{-5} = (-1)^{-5} x^{5(-5)} y^{-5(-5)} = \frac{1}{(-1)^5}x^{-25} y^{25} = -\frac{y^{25}}{x^{25}}$

6.6 Distributing with Variables

1) $2(x + 4) = 2(x) + 2(4) = 2x + 8$

2) $x(3 - x) = x(3) + x(-x) = 3x - x^2$

3) $3x(x - 2) = 3x(x) + 3x(-2) = 3x^2 - 6x$

4) $5x(2x + 3) = 5x(2x) + 5x(3) = 10x^2 + 15x$

5) $4x(-x^2 - 5) = 4x(-x^2) + 4x(-5) = -4x^3 - 20x$

6) $2x(3x^2 + 4x) = 2x(3x^2) + 2x(4x) = 6x^3 + 8x^2$

7) $3x^2(4x^2 - 3) = 3x^2(4x^2) + 3x^2(-3) = 12x^4 - 9x^2$

8) $x(x^2 - x + 1) = x(x^2) + x(-x) + x(1) = x^3 - x^2 + x$

9) $3x(2x^2 + 5x - 7) = 3x(2x^2) + 3x(5x) + 3x(-7) = 6x^3 + 15x^2 - 21x$

10) $4x^2(3x^7 - 6x^4 + 9) = 4x^2(3x^7) + 4x^2(-6x^4) + 4x^2(9) = 12x^9 - 24x^6 + 36x^2$

11) $8x^4(-7x^6 - 5x^5 + 4x^4) = 8x^4(-7x^6) + 8x^4(-5x^5) + 8x^4(4x^4) = -56x^{10} - 40x^9 + 32x^8$

12) $-7(6x + 8) = -7(6x) - 7(8) = -42x - 56$

13) $-x(4x - 5) = -x(4x) - x(-5) = -4x^2 + 5x$

14) $-(7 - x) = -(7) - (-x) = -7 + x$ Alternate: $x - 7$

15) $-9x(-8x^2 - 6x) = -9x(-8x^2) - 9x(-6x) = 72x^3 + 54x^2$

16) $-6x^2(-7x + 6) = -6x^2(-7x) - 6x^2(6) = 42x^3 - 36x^2$

17) $-8x(8x^2 + 6x) = -8x(8x^2) + 8x(6x) = -64x^3 - 48x^2$

18) $-5(x^2 - 4x + 9) = -5(x^2) - 5(-4x) - 5(9) = -5x^2 + 20x - 45$

19) $-(2x^4 - 4x + 8) = -(2x^4) - (-4x) - (8) = -2x^4 + 4x - 8$

20) $-x(-5x^2 + 6x - 3) = -x(-5x^2) - x(6x) - x(-3) = 5x^3 - 6x^2 + 3x$

21) $-7x^2(8x^5 - 6x^3 + 4x) = -7x^2(8x^5) - 7x^2(-6x^3) - 7x^2(4x) = -56x^7 + 42x^5 - 28x^3$

22) $-9x^5(7x^8 + 5x^6 - 3x^4) = -9x^5(7x^8) - 9x^5(5x^6) - 9x^5(-3x^4) = -63x^{13} - 45x^{11} + 27x^9$

6.7 The FOIL Method

1) $(x + 4)(x + 3) = x(x) + x(3) + 4(x) + 4(3) = x^2 + 3x + 4x + 12 = x^2 + 7x + 12$

2) $(x - 2)(x + 5) = x(x) + x(5) - 2(x) - 2(5) = x^2 + 5x - 2x - 10 = x^2 + 3x - 10$

3) $(x - 1)(x - 2) = x(x) + x(-2) - 1(x) - 1(-2) = x^2 - 2x - 1x + 2 = x^2 - 3x + 2$

4) $(-x + 6)(x + 4) = -x(x) - x(4) + 6(x) + 6(4) = -x^2 - 4x + 6x + 24 = -x^2 + 2x + 24$

5) $(3 - x)(8 - x) = 3(8) + 3(-x) - x(8) - x(-x) = 24 - 3x - 8x + x^2 = x^2 - 11x + 24$

6) $(-x - 5)(x + 6) = -x(x) - x(6) - 5(x) - 5(6) = -x^2 - 6x - 5x - 30 = -x^2 - 11x - 30$

7) $(x^2 + x)(x + 1) = x^2(x) + x^2(1) + x(x) + x(1) = x^3 + x^2 + x^2 + x = x^3 + 2x^2 + x$

8) $(-x - 7)(-x - 8) = -x(-x) - x(-8) - 7(-x) - 7(-8) = x^2 + 8x + 7x + 56$
$= x^2 + 15x + 56$

9) $(3x + 2)(4x + 5) = 3x(4x) + 3x(5) + 2(4x) + 2(5) = 12x^2 + 15x + 8x + 10$
$= 12x^2 + 23x + 10$

10) $(x - 6)(-x - 9) = x(-x) + x(-9) - 6(-x) - 6(-9) = -x^2 - 9x + 6x + 54$
$= -x^2 - 3x + 54$

11) $(2x^2 + 4)(3x^2 - 5) = 2x^2(3x^2) + 2x^2(-5) + 4(3x^2) + 4(-5)$
$= 6x^4 - 10x^2 + 12x^2 - 20 = 6x^4 + 2x^2 - 20$

12) $(5x^2 - 3)(2x + 6) = 5x^2(2x) + 5x^2(6) - 3(2x) - 3(6) = 10x^3 + 30x^2 - 6x - 18$

13) $(x^2 + 1)(x^2 - 1) = x^2(x^2) + x^2(-1) + 1(x^2) + 1(-1) = x^4 - x^2 + x^2 - 1 = x^4 - 1$

14) $(4x^3 + 2x)(3x^2 - 2x) = 4x^3(3x^2) + 4x^3(-2x) + 2x(3x^2) + 2x(-2x)$
$= 12x^5 - 8x^4 + 6x^3 - 4x^2$

15) $(6x^2 - 7x)(5x - 3) = 6x^2(5x) + 6x^2(-3) - 7x(5x) - 7x(-3)$
$= 30x^3 - 18x^2 - 35x^2 + 21x = 30x^3 - 53x^2 + 21x$

16) $(4x + 7)(4x + 7) = 4x(4x) + 4x(7) + 7(4x) + 7(7) = 16x^2 + 28x + 28x + 49$
$= 16x^2 + 56x + 49$

17) $(8x^3 + 6x)(7x^4 - 5x^2) = 8x^3(7x^4) + 8x^3(-5x^2) + 6x(7x^4) + 6x(-5x^2)$
$= 56x^7 - 40x^5 + 42x^5 - 30x^3 = 56x^7 + 2x^5 - 30x^3$

6.8 Distributing More Than Two Terms

1) $(x + 2)(x^2 + 4x + 3) = x(x^2) + x(4x) + x(3) + 2(x^2) + 2(4x) + 2(3)$
$= x^3 + 4x^2 + 3x + 2x^2 + 8x + 6 = x^3 + (4 + 2)x^2 + (3 + 8)x + 6 = x^3 + 6x^2 + 11x + 6$

2) $(2x - 1)(x^2 - 3x + 5) = 2x(x^2) + 2x(-3x) + 2x(5) - 1(x^2) - 1(-3x) - 1(5)$
$= 2x^3 - 6x^2 + 10x - 1x^2 + 3x - 5 = 2x^3 - 7x^2 + 13x - 5$

3) $(x^2 - 3x + 5)(x^2 + 2x - 4)$
$= x^2(x^2) + x^2(2x) + x^2(-4) - 3x(x^2) - 3x(2x) - 3x(-4) + 5(x^2) + 5(2x) + 5(-4)$
$= x^4 + 2x^3 - 4x^2 - 3x^3 - 6x^2 + 12x + 5x^2 + 10x - 20 = x^4 - x^3 - 5x^2 + 22x - 20$

4) $(2x^2 - x + 4)(3x^2 + 5x - 7)$
$= 2x^2(3x^2) + 2x^2(5x) + 2x^2(-7) - x(3x^2) - x(5x) - x(-7) + 4(3x^2) + 4(5x) + 4(-7)$
$= 6x^4 + 10x^3 - 14x^2 - 3x^3 - 5x^2 + 7x + 12x^2 + 20x - 28 = 6x^4 + 7x^3 - 7x^2 + 27x - 28$

6.9 Expansion of Powers

1) $(x + 4)^2 = (x + 4)(x + 4) = x(x) + x(4) + 4(x) + 4(4) = x^2 + 4x + 4x + 16$
$= x^2 + (4 + 4)x + 16 = x^2 + 8x + 16$

2) $(3x - 5)^2 = (3x - 5)(3x - 5)$
$= 3x(3x) + 3x(-5) - 5(3x) - 5(-5) = 9x^2 - 15x - 15x + 25 = 9x^2 - 30x + 25$

3) $(x + 2)^3 = (x + 2)(x + 2)(x + 2) = (x^2 + 2x + 2x + 4)(x + 2) = (x^2 + 4x + 4)(x + 2)$
$= x^2(x) + x^2(2) + 4x(x) + 4x(2) + 4(x) + 4(2) = x^3 + 2x^2 + 4x^2 + 8x + 4x + 8$
$= x^3 + 6x^2 + 12x + 8$

4) $(2x - 3)^3 = (2x - 3)(2x - 3)(2x - 3) = (4x^2 - 6x - 6x + 9)(2x - 3)$
$= (4x^2 - 12x + 9)(2x - 3)$
$= 4x^2(2x) + 4x^2(-3) - 12x(2x) - 12x(-3) + 9(2x) + 9(-3)$
$= 8x^3 - 12x^2 - 24x^2 + 36x + 18x - 27 = 8x^3 - 36x^2 + 54x - 27$

6.10 Factoring with Variables

1) $4x^3 + 8x = 4x(x^2 + 2)$

2) $18x^7 - 12x^3 = 6x^3(3x^4 - 2)$

3) $5x^4 - 7x^2 = x^2(5x^2 - 7)$

4) $-x^4 - x^2 = -x^2(x^2 + 1)$

5) $15x + 25 = 5(3x + 5)$

6) $18x^5 - 24x^2 = 6x^2(3x^3 - 4)$

7) $-4x^2 + 6x = -2x(2x - 3)$ Alternative: $2x(-2x + 3)$

8) $48x^9 - 36x^7 = 12x^7(4x^2 - 3)$

9) $21x^5 + 28x = 7x(3x^4 + 4)$

10) $-7x^8 - 11x^5 = -x^5(7x^3 + 11)$

11) $12x^5 + 20x^4 - 16x^3 = 4x^3(3x^2 + 5x - 4)$

12) $15x^4 - 9x^3 + 6x^2 = 3x^2(5x^2 - 3x + 2)$

13) $-8x^3 + 10x^2 - 6x = -2x(4x^2 - 5x + 3)$ Alternative: $2x(-4x^2 + 5x - 3)$

14) $-x^8 - 3x^7 - 2x^5 = -x^5(x^3 + 3x^2 + 2)$

15) $3x^2 - 6x + 9 = 3(x^2 - 2x + 3)$

16) $36x^9 + 27x^8 - 45x^7 = 9x^7(4x^2 + 3x - 5)$

17) $48x^5 + 24x^4 + 60x^3 = 12x^3(4x^2 + 2x + 5)$

18) $-45x^6 - 30x^5 + 60x^4 = -15x^4(3x^2 + 2x - 4)$ Alternative: $15x^4(-3x^2 - 2x + 4)$

19) $33x^3 - 22x^2 - 55x = 11x(3x^2 - 2x - 5)$ Alternative: $11x(3x - 5)(x + 1)$

20) $-32x^7 - 48x^6 - 24x^3 = -8x^3(4x^4 + 6x^3 + 3)$

21) $12x^6 - 18x^5 + 9x^4 = 3x^4(4x^2 - 6x + 3)$

6.11 The Sum or Difference of Squares

1) $(x + 8)(x + 8) = x^2 + 2x(8) + 8^2 = x^2 + 16x + 64$

2) $(x + 5)(x - 5) = x^2 - 5^2 = x^2 - 25$

3) $(x + 4)^2 = (x + 4)(x + 4) = x^2 + 2(x)(4) + 4^2 = x^2 + 8x + 16$

4) $(x - 9)(x + 9) = (x + 9)(x - 9) = x^2 - 9^2 = x^2 - 81$

5) $(3x + 4)(3x + 4) = (3x)^2 + 2(3x)(4) + 4^2 = 9x^2 + 24x + 16$

6) $(7x + 2)(7x - 2) = (7x)^2 - 2^2 = 49x^2 - 4$

7) $(8x - 3)(8x - 3) = (8x)^2 + 2(8x)(-3) + (-3)^2 = 64x^2 - 48x + 9$

8) $(4x - 5)(4x + 5) = (4x)^2 - 5^2 = 16x^2 - 25$

9) $(9x - 4)^2 = (9x - 4)(9x - 4) = (9x)^2 + 2(9x)(-4) + (-4)^2 = 81x^2 - 72x + 16$

10) $(x^2 + 9)(x^2 - 9) = (x^2)^2 - 9^2 = x^4 - 81$

11) $(x^2 + 6)(x^2 + 6) = (x^2)^2 + 2(x^2)(6) + 6^2 = x^4 + 12x^2 + 36$

12) $(x^3 - 3)(x^3 + 3) = (x^3)^2 - 3^2 = x^6 - 9$

13) $(-x - 1)(x + 1) = -(x + 1)(x + 1) = -(x^2 + 2x + 1) = -x^2 - 2x - 1$

Note: First factor $-x - 1$ as $-(x + 1)$ according to Sec. 6.10.

14) $x^2 - 9 = x^2 - 3^2 = (x + 3)(x - 3)$

15) $x^2 + 4x + 4 = x^2 + 2x2 + 2^2 = (x + 2)^2$

16) $4x^2 - 64 = (2x)^2 - 8^2 = (2x + 8)(2x - 8)$ Alternative: $4(x + 4)(x - 4)$

17) $x^2 + 18x + 81 = x^2 + 2(x)(9) + 9^2 = (x + 9)^2$

18) $x^2 - 1 = x^2 - 1^2 = (x + 1)(x - 1)$

19) $16x^2 + 48x + 36 = (4x)^2 + 2(4x)(6) + 6^2 = (4x + 6)^2$ Alternative: $4(2x + 3)^2$

20) $49x^2 - 100 = (7x)^2 - 10^2 = (7x + 10)(7x - 10)$

6.12 Square Roots with Variables

1) $\sqrt{9x} = \sqrt{9}\sqrt{x} = 3\sqrt{x}$

2) $\sqrt{5x^2} = \sqrt{5}\sqrt{x^2} = x\sqrt{5}$

3) $\sqrt{11x^2} = \sqrt{11}\sqrt{x^2} = x\sqrt{11}$

4) $\sqrt{18x} = \sqrt{9}\sqrt{2x} = 3\sqrt{2x}$

5) $\sqrt{6x^2} = \sqrt{6}\sqrt{x^2} = x\sqrt{6}$

6) $\sqrt{36x} = \sqrt{36}\sqrt{x} = 6\sqrt{x}$

7) $\sqrt{20x} = \sqrt{4}\sqrt{5x} = 2\sqrt{5x}$

8) $\sqrt{49x^2} = \sqrt{49}\sqrt{x^2} = 7x$

9) $\sqrt{x^4} = \sqrt{x^2}\sqrt{x^2} = x^2$

10) $\sqrt{81x} = \sqrt{81}\sqrt{x} = 9\sqrt{x}$

11) $\sqrt{8x^2} = \sqrt{4x^2}\sqrt{2} = 2x\sqrt{2}$

12) $\sqrt{x^3} = \sqrt{x^2}\sqrt{x^1} = x\sqrt{x}$

13) $\sqrt{x^{10}} = \sqrt{x^5}\sqrt{x^5} = x^5$

14) $\sqrt{x^5} = \sqrt{x^4}\sqrt{x^1} = x^2\sqrt{x}$

15) $\frac{1}{\sqrt{3}} = \frac{1}{\sqrt{3}}\frac{\sqrt{3}}{\sqrt{3}} = \frac{\sqrt{3}}{3}$

16) $\frac{1}{\sqrt{2x}} = \frac{1}{\sqrt{2x}}\frac{\sqrt{2x}}{\sqrt{2x}} = \frac{\sqrt{2x}}{2x}$

17) $\frac{x}{\sqrt{5}} = \frac{x}{\sqrt{5}}\frac{\sqrt{5}}{\sqrt{5}} = \frac{x\sqrt{5}}{5}$

Note: In this context it is common to work only with the positive roots.

18) $\frac{3}{\sqrt{x}} = \frac{3}{\sqrt{x}}\frac{\sqrt{x}}{\sqrt{x}} = \frac{3\sqrt{x}}{x}$

19) $\frac{\sqrt{2}}{\sqrt{x}} = \frac{\sqrt{2}}{\sqrt{x}}\frac{\sqrt{x}}{\sqrt{x}} = \frac{\sqrt{2x}}{x}$

20) $\dfrac{\sqrt{x}}{\sqrt{3}} = \dfrac{\sqrt{x}}{\sqrt{3}} \dfrac{\sqrt{3}}{\sqrt{3}} = \dfrac{\sqrt{3x}}{3}$

21) $\dfrac{6}{\sqrt{2x}} = \dfrac{6}{\sqrt{2x}} \dfrac{\sqrt{2x}}{\sqrt{2x}} = \dfrac{6\sqrt{2x}}{2x} = \dfrac{3\sqrt{2x}}{x}$

22) $\dfrac{\sqrt{2}}{\sqrt{3}} = \dfrac{\sqrt{2}}{\sqrt{3}} \dfrac{\sqrt{3}}{\sqrt{3}} = \dfrac{\sqrt{6}}{3}$

23) $\dfrac{1}{x\sqrt{x}} = \dfrac{1}{x\sqrt{x}} \dfrac{\sqrt{x}}{\sqrt{x}} = \dfrac{\sqrt{x}}{xx} = \dfrac{\sqrt{x}}{x^2}$ since $xx = x^1 x^1 = x^2$

24) $\dfrac{x}{\sqrt{x}} = \dfrac{x}{\sqrt{x}} \dfrac{\sqrt{x}}{\sqrt{x}} = \dfrac{x\sqrt{x}}{x} = \sqrt{x}$

25) $\dfrac{6}{\sqrt{2}} = \dfrac{6}{\sqrt{2}} \dfrac{\sqrt{2}}{\sqrt{2}} = \dfrac{6\sqrt{2}}{2} = 3\sqrt{2}$

26) $\dfrac{1}{2\sqrt{2}} = \dfrac{1}{2\sqrt{2}} \dfrac{\sqrt{2}}{\sqrt{2}} = \dfrac{\sqrt{2}}{2(2)} = \dfrac{\sqrt{2}}{4}$

27) $\dfrac{x^2}{\sqrt{x}} = \dfrac{x^2}{\sqrt{x}} \dfrac{\sqrt{x}}{\sqrt{x}} = \dfrac{x^2\sqrt{x}}{x} = x\sqrt{x}$

28) $\dfrac{6}{\sqrt{3}} = \dfrac{6}{\sqrt{3}} \dfrac{\sqrt{3}}{\sqrt{3}} = \dfrac{6\sqrt{3}}{3} = 2\sqrt{3}$

29) $\dfrac{10x}{\sqrt{5}} = \dfrac{10x}{\sqrt{5}} \dfrac{\sqrt{5}}{\sqrt{5}} = \dfrac{10x\sqrt{5}}{5} = 2x\sqrt{5}$

30) $\dfrac{x}{\sqrt{7x}} = \dfrac{x}{\sqrt{7x}} \dfrac{\sqrt{7x}}{\sqrt{7x}} = \dfrac{x\sqrt{7x}}{7x} = \dfrac{\sqrt{7x}}{7}$

31) $\dfrac{14}{\sqrt{7x}} = \dfrac{14}{\sqrt{7x}} \dfrac{\sqrt{7x}}{\sqrt{7x}} = \dfrac{14\sqrt{7x}}{7x} = \dfrac{2\sqrt{7x}}{x}$

7 Solving Equations

7.1 One-Step Equations

1) $x = 7$ Check: $7 + 8 = 15$

2) $x = 11$ Check: $11 - 7 = 4$

3) $x = 9$ Check: $7(9) = 63$

4) $x = 54$ Check: $\frac{54}{6} = 9$

5) $x = 10$ Check: $10 - 5 = 5$

6) $x = 9$ Check: $9(9) = 81$

7) $x = 24$ Check: $\frac{24}{4} = 6$

8) $x = 3$ Check: $3 + 9 = 12$

9) $x = 13$ Check: $4(13) = 52$

10) $x = 9$ Check: $9 - 8 = 1$

11) $x = 6$ Check: $6 + 5 = 11$

12) $x = 20$ Check: $\frac{20}{2} = 10$

13) $x = 16$ Check: $16 - 6 = 10$

14) $x = 7$ Check: $6(7) = 42$

15) $x = 49$ Check: $\frac{49}{7} = 7$

16) $x = 6$ Check: $6 + 3 = 9$

17) $x = 13$ Check: $13 - 4 = 9$

18) $x = 27$ Check: $\frac{27}{3} = 9$

19) $x = 10$ Check: $10(10) = 100$

20) $x = 8$ Check: $8 + 7 = 15$

21) $x = 22$ Check: $22 - 9 = 13$

22) $x = 32$ Check: $\frac{32}{8} = 4$

23) $x = 4$ Check: $4 + 6 = 10$

24) $x = 12$ Check: $5(12) = 60$

25) $x = 27$ Check: $\frac{27}{9} = 3$

26) $x = 15$ Check: $15 - 3 = 12$

27) $x = 7$ Check: $8(7) = 56$

28) $x = 10$ Check: $10 + 4 = 14$

29) $x = 30$ Check: $30 - 10 = 20$

30) $x = 60$ Check: $\frac{60}{10} = 6$

7.2 Isolate the Unknown

1) $x = 7$ Check: $2x + 3 = 2(7) + 3 = 14 + 3 = 17$

2) $x = 6$ Check: $3x + 24 = 3(6) + 24 = 18 + 24 = 42$ agrees with $7x = 7(6) = 42$

3) $x = 4$ Check: $5x = 5(4) = 20$ agrees with $12 + 2x = 12 + 2(4) = 12 + 8 = 20$

4) $x = 6$ Check: $29 = 4x + 5 = 4(6) + 5 = 24 + 5 = 29$

5) $x = 5$ Check: $8x - 9 = 8(5) - 9 = 40 - 9 = 31$

6) $x = 5$ Check: $15 - x = 15 - 5 = 10$ agrees with $2x = 2(5) = 10$

7) $x = 8$ Check: $6x = 6(8) = 48$ agrees with $72 - 3x = 72 - 3(8) = 72 - 24 = 48$

8) $x = 7$ Check: $27 = 5x - 8 = 5(7) - 8 = 35 - 8 = 27$

9) $x = 7$ Check: $7 + x = 7 + 7 = 14$ agrees with $2x = 2(7) = 14$

10) $x = 1$ Check: $5 + 3x = 5 + 3(1) = 5 + 3 = 8$ agrees with $8x = 8(1) = 8$

11) $x = 5$ Check: $-3 + 4x = -3 + 4(5) = -3 + 20 = 17$

12) $x = 10$ Check: $16 = -4 + 2x = -4 + 2(10) = -4 + 20 = 16$

13) $x = 6$ Check: $5x - 2 = 5(6) - 2 = 30 - 2 = 28$ agrees with $3x + 10 = 3(6) + 10 = 18 + 10 = 28$

14) $x = 4$ Check: $24 - x = 24 - 4 = 20$ agrees with $3x + 8 = 3(4) + 8 = 12 + 8 = 20$

15) $x = 7$ Check: $2 + 5x = 2 + 5(7) = 2 + 35 = 37$ agrees with $9 + 4x = 9 + 4(7) = 9 + 28 = 37$

16) $x = 9$ Check: $3x + 38 = 3(9) + 38 = 27 + 38 = 65$ agrees with $8x - 7 = 8(9) - 7 = 72 - 7 = 65$

17) $x = 8$ Check: $6x - 4 = 6(8) - 4 = 48 - 4 = 44$ agrees with
$9x - 28 = 9(8) - 28 = 72 - 28 = 44$

18) $x = 2$ Check: $11 + 2x = 11 + 2(2) = 11 + 4 = 15$ agrees with $6x + 3 = 6(2) + 3 = 12 + 3 = 15$

19) $x = 6$ Check: $15 - x = 15 - 6 = 9$ agrees with $3 + x = 3 + 6 = 9$

20) $x = 2$ Check: $5 - 3x = 5 - 3(2) = 5 - 6 = -1$ agrees with $17 - 9x = 17 - 9(2) = 17 - 18 = -1$

21) $x = 3$ Check: $9 + 2x = 9 + 2(3) = 9 + 6 = 15$ agrees with $8x - 3x = 8(3) - 3(3) = 24 - 9 = 15$

22) $x = 4$ Check: $7 + 6x = 7 + 6(4) = 7 + 24 = 31$ agrees with $36 - 5 = 31$

23) $x = 7$ Check: $64 - 4x = 64 - 4(7) = 64 - 28 = 36$ agrees with $8 + 4x = 8 + 4(7) = 8 + 28 = 36$

24) $x = 2$ Check: $-2x - 3 = -2(2) - 3 = -4 - 3 = -7$ agrees with $-7x + 7 = -7(2) + 7 = -14 + 7 = -7$

7.3 Variables with Negative Signs

1) $x = -3$ Check: $7 - 2x = 7 - 2(-3) = 7 + 6 = 13$

2) $x = -4$ Check: $7x = 7(-4) = -28$ agrees with $4x - 12 = 4(-4) - 12 = -16 - 12 = -28$

3) $x = -7$ Check: $9x = 9(-7) = -63$ agrees with $-35 + 4x = -35 + 4(-7) = -35 - 28 = -63$

4) $x = 2$ Check: $12 - 5x = 12 - 5(2) = 12 - 10 = 2$

5) $x = -4$ Check: $28 + 6x = 28 + 6(-4) = 28 - 24 = 4$

6) $x = 3$ Check: $-18 + 3x = -18 + 3(3) = -18 + 9 = -9$ agrees with $-3x = -3(3) = -9$

7) $x = -7$ Check: $5x + 14 = 5(-7) + 14 = -35 + 14 = -21$ agrees with $3x = 3(-7) = -21$

8) $x = 9$ Check: $40 - 8x = 40 - 8(9) = 40 - 72 = -32$

9) $x = -8$ Check: $44 = 12 - 4(x) = 12 - 4(-8) = 12 - (-32) = 12 + 32 = 44$

10) $x = -5$ Check: $-7x = -7(-5) = 35$ agrees with $20 - 3x = 20 - 3(-5) = 20 - (-15) = 35$

11) $x = 8$ Check: $2x = 2(8) = 16$ agrees with $9x - 56 = 9(8) - 56 = 72 - 56 = 16$

12) $x = 2$ Check: $-1 = 2x - 5 = 2(2) - 5 = 4 - 5 = -1$

13) $x = -4$ Check: $-x = -(-4) = 4$

14) $x = 3$ Check: $1 - x = 1 - 3 = -2$

15) $x = 4$ Check: $-x - 4 = -4 - 4 = -8$

16) $x = 9$ Check: $-x = -9$

17) $x = -3$ Check: $2x + 3 = 2(-3) + 3 = -6 + 3 = -3$

18) $x = 1$ Check: $6 - x = 6 - 1 = 5$

7.4 Variables with Fractions

1) $x = \frac{4}{5}$ Check: $5x - 1 = 5\left(\frac{4}{5}\right) - 1 = 4 - 1 = 3$

2) $x = \frac{4}{3}$ Check: $3 + 9x = 3 + 9\left(\frac{4}{3}\right) = 3 + \frac{36}{3} = 3 + 12 = 15$

3) $x = \frac{2}{3}$ Check: $3x = 3\left(\frac{2}{3}\right) = 2$

4) $x = \frac{4}{9}$ Check: $2x = 2\left(\frac{4}{9}\right) = \frac{8}{9}$

5) $x = \frac{27}{8}$ Check: $x - \frac{7}{8} = \frac{27}{8} - \frac{7}{8} = \frac{20}{8} = \frac{20 \div 4}{8 \div 4} = \frac{5}{2}$

6) $x = \frac{1}{6}$ Check: $3x + \frac{1}{3} = 3\left(\frac{1}{6}\right) + \frac{1}{3} = \frac{3}{6} + \frac{1}{3} = \frac{3}{6} + \frac{1 \cdot 2}{3 \cdot 2} = \frac{3}{6} + \frac{2}{6} = \frac{5}{6}$

7) $x = -\frac{5}{3}$ Check: $10 + 2x = 10 + 2\left(-\frac{5}{3}\right) = 10 - \frac{10}{3} = \frac{30}{3} - \frac{10}{3} = \frac{30 - 10}{3} = \frac{20}{3}$ agrees with

$-4x = -4\left(-\frac{5}{3}\right) = \frac{20}{3}$

8) $x = \frac{7}{9}$ Check: $5 - 9x = 5 - 9\left(\frac{7}{9}\right) = 5 - 7 = -2$

9) $x = -\frac{3}{4}$ Check: $4x = 4\left(-\frac{3}{4}\right) = -3$

10) $x = \frac{4}{7}$ Check: $4 - 7x = 4 - 7\left(\frac{4}{7}\right) = 4 - 4 = 0$

11) $x = \frac{6}{5}$ Check: $\frac{4}{5} + x = \frac{4}{5} + \frac{6}{5} = \frac{10}{5} = 2$

12) $x = \frac{17}{24}$ Check: $2x - \frac{2}{3} = 2\left(\frac{17}{24}\right) - \frac{2}{3} = \frac{17}{12} - \frac{2}{3} = \frac{17}{12} - \frac{2 \cdot 4}{3 \cdot 4} = \frac{17}{12} - \frac{8}{12} = \frac{9}{12} = \frac{9 \div 3}{12 \div 3} = \frac{3}{4}$

7.5 Roots and Exponents of Variables

1) $x = 36$ Check: $\sqrt{x} = \sqrt{36} = 6$

2) $x = \pm 7$ Check: $x^2 = (-7)^2 = 49$ and $x^2 = 7^2 = 49$

3) $x = 4$ Check: $x^3 = 4^3 = 64$

4) $x = 64$ Check: $\sqrt{x} = \sqrt{64} = -8$ or 8

5) $x = 49$ Check: $\sqrt{x} - 2 = \sqrt{49} - 2 = 7 - 2 = 5$

6) $x = \pm 4$ Check: $4x^4 = 4(-4)^4 = 4(256) = 1024$ and $4x^4 = 4(4)^4 = 4(256) = 1024$

7) $x = \pm 5$ Check: $2x^2 = 2(-5)^2 = 2(25) = 50$ and $2x^2 = 2(5)^2 = 2(25) = 50$

8) $x = 16$ Check: $3\sqrt{x} = 3\sqrt{16} = 3(4) = 12$

9) $x = 4$ Check: $2 - \sqrt{x} = 2 - \sqrt{4} = 2 - 2 = 0$

10) $x = \pm 5$ Check: $x^2 - 9 = 5^2 - 9 = 25 - 9 = 16$ and $x^2 - 9 = (-5)^2 - 9 = 25 - 9 = 16$

11) $x = -5$ Check: $25 - x^3 = 25 - (-5)^3 = 25 - (-125) = 25 + 125 = 150$

12) $x = 81$ Check: $8\sqrt{x} = 8\sqrt{81} = 8(9) = 72$

7.6 Variables in a Denominator

1) $x = 18$ Check: $\frac{1}{x} + \frac{1}{9} = \frac{1}{18} + \frac{1}{9} = \frac{1}{18} + \frac{1 \cdot 2}{9 \cdot 2} = \frac{1}{18} + \frac{2}{18} = \frac{3}{18} = \frac{3 \div 3}{18 \div 3} = \frac{1}{6}$

2) $x = 3$ Check: $\frac{5}{4} - \frac{1}{x} = \frac{5}{4} - \frac{1}{3} = \frac{5 \cdot 3}{4 \cdot 3} - \frac{1 \cdot 4}{3 \cdot 4} = \frac{15}{12} - \frac{4}{12} = \frac{15-4}{12} = \frac{11}{12}$

3) $x = \frac{5}{4}$ Check: $\frac{5}{3} - \frac{1}{x} = \frac{5}{3} - \frac{1}{\frac{5}{4}} = \frac{5}{3} - \frac{4}{5} = \frac{5 \cdot 5}{3 \cdot 5} - \frac{4 \cdot 3}{5 \cdot 3} = \frac{25}{15} - \frac{12}{15} = \frac{25-12}{15} = \frac{13}{15}$

4) $x = 24$ Check: $\frac{1}{x} + \frac{5}{6} = \frac{1}{24} + \frac{5}{6} = \frac{1}{24} + \frac{5 \cdot 4}{6 \cdot 4} = \frac{1}{24} + \frac{20}{24} = \frac{1+20}{24} = \frac{21}{24} = \frac{21 \div 3}{24 \div 3} = \frac{7}{8}$

7.7 Cross Multiplying

1) $x = 3$ Check: $\frac{x}{4} = \frac{3}{4} = \frac{3 \cdot 4}{4 \cdot 4} = \frac{12}{16}$

2) $x = 3$ Check: $\frac{2}{x} = \frac{2}{3} = \frac{2 \cdot 4}{3 \cdot 4} = \frac{8}{12}$

3) $x = \frac{3}{2}$ Check: $\frac{x}{2} = \frac{3}{2} \div 2 = \frac{3}{2} \div \frac{2}{1} = \frac{3}{2} \cdot \frac{1}{2} = \frac{3 \cdot 1}{2 \cdot 2} = \frac{3}{4}$

4) $x = \frac{15}{8}$ Check: $\frac{5}{x} = 5 \div \frac{15}{8} = \frac{5}{1} \div \frac{15}{8} = \frac{5}{1} \cdot \frac{8}{15} = \frac{5 \cdot 8}{1 \cdot 15} = \frac{40}{15} = \frac{40 \div 5}{15 \div 5} = \frac{8}{3}$

5) $x = \frac{9}{4}$ Check: $\frac{x}{6} = \frac{9}{4} \div 6 = \frac{9}{4} \div \frac{6}{1} = \frac{9}{4} \cdot \frac{1}{6} = \frac{9 \cdot 1}{4 \cdot 6} = \frac{9}{24} = \frac{9 \div 3}{24 \div 3} = \frac{3}{8}$

6) $x = \frac{2}{3}$ Check: $\frac{5}{x} = 5 \div \frac{2}{3} = \frac{5}{1} \div \frac{2}{3} = \frac{5}{1} \cdot \frac{3}{2} = \frac{5 \cdot 3}{1 \cdot 2} = \frac{15}{2}$

7) $x = \frac{2}{3}$ Check: $\frac{x}{7} = \frac{2}{3} \div 7 = \frac{2}{3} \div \frac{7}{1} = \frac{2}{3} \cdot \frac{1}{7} = \frac{2 \cdot 1}{3 \cdot 7} = \frac{2}{21}$

8) $x = 8$ Check: $\frac{10}{x} = \frac{10}{8} = \frac{10 \div 2}{8 \div 2} = \frac{5}{4}$

9) $x = \frac{1}{3}$ Check: $\frac{3}{2x} = \frac{3}{2\left(\frac{1}{3}\right)} = \frac{3}{\frac{2}{3}} = 3 \div \frac{2}{3} = \frac{3}{1} \div \frac{2}{3} = \frac{3}{1} \cdot \frac{3}{2} = \frac{3 \cdot 3}{1 \cdot 2} = \frac{9}{2}$

10) $x = \frac{5}{2}$ Check: $\frac{3x}{2} = \frac{3\left(\frac{5}{2}\right)}{2} = \frac{\frac{15}{2}}{2} = \frac{15}{2} \div 2 = \frac{15}{2} \div \frac{2}{1} = \frac{15}{2} \cdot \frac{1}{2} = \frac{15 \cdot 1}{2 \cdot 2} = \frac{15}{4}$

11) $x = \frac{1}{6}$ Check: $\frac{5}{8x} = \frac{5}{8\left(\frac{1}{6}\right)} = \frac{5}{\frac{8}{6}} = 5 \div \frac{8}{6} = \frac{5}{1} \div \frac{8}{6} = \frac{5}{1} \cdot \frac{6}{8} = \frac{5 \cdot 6}{1 \cdot 8} = \frac{30}{8} = \frac{30 \div 2}{8 \div 2} = \frac{15}{4}$

12) $x = \frac{5}{4}$ Check: $\frac{4x}{35} = \frac{4\left(\frac{5}{4}\right)}{35} = \frac{5}{35} = \frac{5 \div 5}{35 \div 5} = \frac{1}{7}$

13) $x = 1$ Check: $\frac{6x}{8} = \frac{6(1)}{8} = \frac{6}{8} = \frac{6 \div 2}{8 \div 2} = \frac{3}{4}$

14) $x = \frac{5}{2}$ Check: $\frac{4}{5x} = \frac{4}{5\left(\frac{5}{2}\right)} = \frac{4}{\frac{25}{2}} = 4 \div \frac{25}{2} = \frac{4}{1} \div \frac{25}{2} = \frac{4}{1} \cdot \frac{2}{25} = \frac{4 \cdot 2}{1 \cdot 25} = \frac{8}{25}$

7.8 Special Solutions

1) all real numbers since $2 = 2$

2) no solution since $2 \neq 3$

3) no solution since $5 \neq 3$

4) all real numbers since $0 = 0$

5) no solution since $0 \neq 1$

6) no solution since $1 \neq 0$

7) all real numbers since $0 = 0$

8) all real numbers since $\frac{1}{2} = \frac{1}{2}$

8 Ratio Problems

8.1 Reducing Ratios

1) $4:16 = \frac{4}{16} = \frac{4 \div 4}{16 \div 4} = \frac{1}{4} = 1{:}4$

2) $12{:}18 = \frac{12}{18} = \frac{12 \div 6}{18 \div 6} = \frac{2}{3} = 2{:}3$

3) $14{:}6 = \frac{14}{6} = \frac{14 \div 2}{6 \div 2} = \frac{7}{3} = 7{:}3$

4) $4{:}18 = \frac{4}{18} = \frac{4 \div 2}{18 \div 2} = \frac{2}{9} = 2{:}9$

5) $12{:}16 = \frac{12}{16} = \frac{12 \div 4}{16 \div 4} = \frac{3}{4} = 3{:}4$

6) $8{:}24 = \frac{8}{24} = \frac{8 \div 8}{24 \div 8} = \frac{1}{3} = 1{:}3$

7) $8{:}2 = \frac{8}{2} = \frac{8 \div 2}{2 \div 2} = \frac{4}{1} = 4{:}1$

8) $14{:}38 = \frac{14}{38} = \frac{14 \div 2}{38 \div 2} = \frac{7}{19} = 7{:}19$

9) $16{:}24 = \frac{16}{24} = \frac{16 \div 8}{24 \div 8} = \frac{2}{3} = 2{:}3$

10) $22{:}34 = \frac{22}{34} = \frac{22 \div 2}{34 \div 2} = \frac{11}{17} = 11{:}17$

11) $40{:}15 = \frac{40}{15} = \frac{40 \div 5}{15 \div 5} = \frac{8}{3} = 8{:}3$

12) $26{:}4 = \frac{26}{4} = \frac{26 \div 2}{4 \div 2} = \frac{13}{2} = 13{:}2$

13) $18{:}24 = \frac{18}{24} = \frac{18 \div 6}{24 \div 6} = \frac{3}{4} = 3{:}4$

14) $21{:}48 = \frac{21}{48} = \frac{21 \div 3}{48 \div 3} = \frac{7}{16} = 7{:}16$

15) $44{:}8 = \frac{44}{8} = \frac{44 \div 4}{8 \div 4} = \frac{11}{2} = 11{:}2$

16) $30{:}35 = \frac{30}{35} = \frac{30 \div 5}{35 \div 5} = \frac{6}{7} = 6{:}7$

17) $34{:}12 = \frac{34}{12} = \frac{34 \div 2}{12 \div 2} = \frac{17}{6} = 17{:}6$

18) $42{:}52 = \frac{42}{52} = \frac{42 \div 2}{52 \div 2} = \frac{21}{26} = 21{:}26$

8.2 The Part and the Whole

1) (a) $\frac{4}{6} = \frac{4 \div 2}{6 \div 2} = \frac{2}{3} = 2{:}3$ (b) $\frac{6}{10} = \frac{6 \div 2}{10 \div 2} = \frac{3}{5} = 3{:}5$ (c) $\frac{10}{4} = \frac{10 \div 2}{4 \div 2} = \frac{5}{2} = 5{:}2$

2) $7{:}4$ (since the whole is $3 + 4 = 7$)

3) (a) $\frac{9}{6} = \frac{9 \div 3}{6 \div 3} = \frac{3}{2} = 3{:}2$ (b) $\frac{6}{3} = \frac{6 \div 3}{3 \div 3} = \frac{2}{1} = 2{:}1$ (c) $\frac{3}{9} = \frac{3 \div 3}{9 \div 3} = \frac{1}{3} = 1{:}3$

4) $1{:}3$ (for pens, use $4 - 3 = 1$)

5) (a) $\frac{40}{32} = \frac{40 \div 8}{32 \div 8} = \frac{5}{4} = 5{:}4$ (b) $\frac{32}{72} = \frac{32 \div 8}{72 \div 8} = \frac{4}{9} = 4{:}9$ (c) $\frac{72}{40} = \frac{72 \div 8}{40 \div 8} = \frac{9}{5} = 9{:}5$

6) $7{:}10$ (since the whole is $3 + 7 = 10$)

7) (a) $\frac{30}{48} = \frac{30 \div 6}{48 \div 6} = \frac{5}{8} = 5{:}8$ (b) $\frac{30}{18} = \frac{30 \div 6}{18 \div 6} = \frac{5}{3} = 5{:}3$

8.3 Ratios and Fractions

1) $\frac{5}{8}$ (since the whole is $5 + 3 = 8$)

2) 2:3 (since the other part is $5 - 2 = 3$)

3) $\frac{3}{20}$ (since the whole is $17 + 3 = 20$)

4) 4:3 (since the other part is $7 - 4 = 3$)

8.4 Ratios and Percents

1) 60% (since the whole is $3 + 2 = 5$ and since $\frac{3}{5} = \frac{3\cdot20}{5\cdot20} = \frac{60}{100} = 60\%$)

2) 3:1 (since $75\% = \frac{75}{100} = \frac{75\div25}{100\div25} = \frac{3}{4}$ and since the other part is $4 - 3 = 1$)

3) 35% (since the other part is $20 - 13 = 7$ and since $\frac{7}{20} = \frac{7\cdot5}{20\cdot5} = \frac{35}{100} = 35\%$)

4) 41:50 (since the other part is $100\% - 18\% = 82\%$ and since $82\% = \frac{82}{100} = \frac{82\div2}{100\div2} = \frac{41}{50}$)

8.5 Unit Ratios

1) $8{:}5 = \frac{8}{5} = \frac{8\cdot2}{5\cdot2} = \frac{16}{10} = 1.6{:}1$

2) $7{:}20 = \frac{7}{20} = \frac{7\cdot5}{20\cdot5} = \frac{35}{100} = 0.35{:}1$

3) $9{:}4 = \frac{9}{4} = \frac{9\cdot25}{4\cdot25} = \frac{225}{100} = 2.25{:}1$

4) $1{:}2 = \frac{1}{2} = \frac{1\cdot5}{2\cdot5} = \frac{5}{10} = 0.5{:}1$

5) $17{:}50 = \frac{17}{50} = \frac{17\cdot2}{50\cdot2} = \frac{34}{100} = 0.34{:}1$

6) $32{:}25 = \frac{32}{25} = \frac{32\cdot4}{25\cdot4} = \frac{128}{100} = 1.28{:}1$

7) $15{:}8 = \frac{15}{8} = \frac{15\cdot125}{8\cdot125} = \frac{1875}{100} = 1.875{:}1$

8) $283{:}100 = \frac{283}{100} = 2.83{:}1$

9) $53{:}10 = \frac{53}{10} = 5.3{:}1$

10) $19{:}200 = \frac{19}{200} = \frac{19\cdot5}{200\cdot5} = \frac{95}{1000} = 0.095{:}1$

11) $7{:}40 = \frac{7}{40} = \frac{7\cdot25}{40\cdot25} = \frac{175}{1000} = 0.175{:}1$

12) $11{:}4 = \frac{11}{4} = \frac{11\cdot25}{4\cdot25} = \frac{275}{100} = 2.75{:}1$

13) $1{:}5 = \frac{1}{5} = \frac{1\cdot2}{5\cdot2} = \frac{2}{10} = 0.2{:}1$

14) $521{:}250 = \frac{521}{250} = \frac{521\cdot4}{250\cdot4} = \frac{2084}{1000} = 2.084{:}1$

15) $999{:}500 = \frac{999}{500} = \frac{999\cdot2}{500\cdot2} = \frac{1998}{1000} = 1.998{:}1$

16) $207{:}125 = \frac{207}{125} = \frac{207\cdot8}{125\cdot8} = \frac{1656}{1000} = 1.656{:}1$

8.6 Linear Proportions

1) 63 girls (7:6 has more girls than boys, and 63 girls is more than 54 boys)

2) 300 desktops (5:2 has more laptops than desktops, and 750 laptops is more than 300 desktops)

3) 90 defective televisions (3:20 has fewer defective televisions than the total number of televisions, and 90 defective televisions is less than 600 televisions sold)

4) 90 didn't graduate (7 out of 10 students graduated, meaning that 3 out of 10 students did not graduate; 210 students graduated, which is more than the 90 who didn't)

5) 63 children (4:9 has fewer parents than children, and 28 parents is less than 63 children)

6) 21 red apples (8:3 has less red apples than the total number of apples, and 21 red apples is less than the total number of 56 apples)

7) 3 buses (12 buses transport 600 students, and it would take 3 buses to transport 150 students)

8) 160 tomatoes are not ripe (7:9 has fewer ripe tomatoes than the total number of tomatoes, and 160 unripe tomatoes is less than the total number of 720 tomatoes)

9 Rate Problems

9.1 Rates and Decimals

1) 81 meters in 20 seconds $= \dfrac{81 \text{ meters}}{20 \text{ seconds}} = \dfrac{81 \times 5 \text{ meters}}{20 \times 5 \text{ seconds}} = \dfrac{405 \text{ meters}}{100 \text{ seconds}} = 4.05$ meters per second

2) 7 pages in 2 minutes $= \dfrac{7 \text{ pages}}{2 \text{ minutes}} = \dfrac{7 \times 5 \text{ pages}}{2 \times 5 \text{ minutes}} = \dfrac{35 \text{ pages}}{10 \text{ minutes}} = 3.5$ pages per minute

3) 1300 calories in 5 servings $= \dfrac{1300 \text{ calories}}{5 \text{ servings}} = \dfrac{1300 \times 2 \text{ calories}}{5 \times 2 \text{ servings}} = \dfrac{2600 \text{ calories}}{10 \text{ servings}} = 260$ calories per serving

4) 4 inches in 25 days $= \dfrac{4 \text{ inches}}{25 \text{ days}} = \dfrac{4 \times 4 \text{ inches}}{25 \times 4 \text{ days}} = \dfrac{16 \text{ inches}}{100 \text{ days}} = 0.16$ inches per day

5) 18 kilometers in 4 hours $= \dfrac{18 \text{ kilometers}}{4 \text{ hours}} = \dfrac{18 \times 25 \text{ kilometers}}{4 \times 25 \text{ hours}} = \dfrac{450 \text{ kilometers}}{100 \text{ hours}} = 4.5$ kilometers per hour

6) 11 grams in 50 milliliters $= \dfrac{11 \text{ grams}}{50 \text{ milliliters}} = \dfrac{11 \times 2 \text{ grams}}{50 \times 2 \text{ milliliters}} = \dfrac{22 \text{ grams}}{100 \text{ milliliters}} = 0.22$ grams per milliliter

7) 87 dollars for 10 tickets $= \dfrac{87 \text{ dollars}}{10 \text{ tickets}} = 8.7$ dollars per ticket

8) 260 miles for 8 gallons $= \dfrac{260 \text{ miles}}{8 \text{ gallons}} = \dfrac{260 \times 125 \text{ miles}}{8 \times 125 \text{ gallons}} = \dfrac{32{,}500 \text{ miles}}{1000 \text{ gallons}} = 32.5$ miles per gallon

9.2 Constant Speed

1) $d = rt = (25 \text{ m/s})(7 \text{ s}) = 175 \text{ m}$

2) $r = \dfrac{d}{t} = \dfrac{600 \text{ ft.}}{5 \text{ s}} = 120 \text{ ft./s}$

3) $t = \dfrac{d}{r} = \dfrac{150 \text{ m}}{3 \text{ m/s}} = 50 \text{ s}$

4) $r = \dfrac{d}{t} = \dfrac{270 \text{ mi.}}{3 \text{ hr.}} = 90 \text{ mph}$

5) $d = rt = (7 \text{ in./min.})(9 \text{ min.}) = 63 \text{ in.}$

6) $t = \dfrac{d}{r} = \dfrac{300 \text{ km}}{50 \text{ km/hr.}} = 6 \text{ hr}$

7) $r = \dfrac{d}{t} = \dfrac{72 \text{ cm}}{9 \text{ s}} = 8 \text{ cm/s}$

9.3 Unit Conversions

1) $8 \text{ ft.} = 8 \text{ ft.} \times \frac{12 \text{ in.}}{1 \text{ ft.}} = 96 \text{ in.}$

2) $21 \text{ ft.} = 21 \text{ ft.} \times \frac{1 \text{ yd.}}{3 \text{ ft.}} = 7 \text{ yd.}$

3) $240 \text{ min.} = 240 \text{ min.} \times \frac{1 \text{ hr.}}{60 \text{ min.}} = 4 \text{ hr.}$

4) $15 \text{ kg} = 15 \text{ kg} \times \frac{1000 \text{ g}}{1 \text{ kg}} = 15{,}000 \text{ g}$

5) $12 \text{ gal.} = 12 \text{ gal.} \times \frac{4 \text{ qt.}}{1 \text{ gal.}} = 48 \text{ qt.}$

6) $96 \text{ hr.} = 96 \text{ hr.} \times \frac{1 \text{ day}}{24 \text{ hr.}} = 4 \text{ days}$

7) $42 \text{ cm} = 42 \text{ cm} \times \frac{10 \text{ mm}}{1 \text{ cm}} = 420 \text{ mm}$

8) $4 \text{ hr.} = 4 \text{ hr.} \times \frac{60 \text{ min.}}{1 \text{ hr.}} \times \frac{60 \text{ s}}{1 \text{ min.}} = 14{,}400 \text{ s}$

9) $252 \text{ in.} = 252 \text{ in.} \times \frac{1 \text{ ft.}}{12 \text{ in.}} \times \frac{1 \text{ yd.}}{3 \text{ ft.}} = 7 \text{ yd.}$

9.4 Converting the Units of Rates

1) $108 \frac{\text{km}}{\text{hr.}} = 108 \frac{\text{km}}{\text{hr.}} \times \frac{1000 \text{ m}}{1 \text{ km}} \times \frac{1 \text{ hr.}}{3600 \text{ s}} = \frac{108 \times 1000}{3600} \frac{\text{m}}{\text{s}} = 30 \text{ m/s}$

2) $8 \frac{\text{qt.}}{\text{s}} = 8 \frac{\text{qt.}}{\text{s}} \times \frac{1 \text{ gal.}}{4 \text{ qt.}} \times \frac{60 \text{ s}}{1 \text{ min.}} = \frac{8 \times 60}{4} \frac{\text{gal.}}{\text{min.}} = 120 \text{ gal./min.}$

9.5 Rate Formulas

1) $I = \frac{V}{R}$

2) $Q = CV$

3) $q = -Mp$

4) $V = \frac{m}{d}$ (because $dV = m$)

5) $b = \frac{d}{ac}$

6) $T = \frac{1}{f}$ (because $fT = 1$)

7) $y = \frac{z}{x}$ (because $xy = z$; note that $1z = z$)

8) $g = \frac{2h}{t^2}$

9.6 Rates

1) rate $= \frac{\text{machines}}{\text{days}}$ Multiply by days to get machines = (rate)(days). Note that 1 week = 7 days.

machines = (rate)(days) = (6)(7) = 42

2) rate $= \frac{\text{pages}}{\text{minutes}}$ Multiply by minutes to get pages = (rate)(minutes). Now divide by the rate

to get minutes $= \frac{\text{pages}}{\text{rate}} = \frac{56}{7} = 8$

3) rate $= \frac{\text{miles}}{\text{gallons}} = \frac{600}{15} = 40$ mpg

4) rate $= \frac{\text{dollars}}{\text{hours}}$ Multiply by hours to get dollars = (rate)(hours). First determine how many

hours are worked in 5 days: hours = $5 \times 8 = 40$ Now multiply the rate ($30 per hour) by the

time (40 hours): dollars = (rate)(hours) = ($30)(40) = $1200

5) rate $= \frac{\text{dollars}}{\text{months}} = \frac{\$750}{10} = \$75$

6) rate $= \frac{\text{gallons}}{\text{minutes}}$ Multiply by minutes to get gallons = (rate)(minutes). Now divide by the rate

to get minutes $= \frac{\text{gallons}}{\text{rate}} = \frac{20}{0.25} = \frac{(20)(4)}{(0.25)(4)} = \frac{80}{1} = 80$

10 Inequalities

10.1 Negative Numbers and Absolute Values

1) $-23 < -19$ (since -23 is more negative than -19, it follows that -23 is less than -19)

2) $-11 > -17$ (since -11 is less negative than -17, it follows that -11 is greater than -17)

3) $0 > -1$ (since 0 is greater than any negative number)

4) $89 < 90$ (since both numbers are positive, this should be very easy)

5) $-101 < -99$ (since -101 is more negative than -99, it follows that -101 is less than -99)

6) $-34 < 26$ (since any negative number is smaller than any positive number)

7) $|-2| > -3$ (since $|-2| = 2$ and $2 > -3$)

8) $13 > |-12|$ (since $|-12| = 12$ and $13 > 12$)

9) $29 < |-31|$ (since $|-31| = 31$ and $29 < 31$)

10) $|-8| < |-9|$ (since $|-8| = 8, |-9| = 9$, and $8 < 9$)

11) $|-33| > |-25|$ (since $|-33| = 33, |-25| = 25$, and $33 > 25$)

12) $|-4| > 0$ (since $|-4| = 4$ and $4 > 0$)

10.2 Comparing Decimals and Percents

1) $0.7 > 0.4$ (since 0 is the same, $.7 > .4$ determines which is greater)

2) $9.5 < 9.6$ (since 9 is the same, $.5 < .6$ determines which is greater)

3) $2.736 > 1.949$ (since $2 > 1$)

4) $3.14 > 3.1$ (since $3.10 = 3.1$ and since $3.14 > 3.10$)

5) $7.234 < 7.237$ (since 7.23 is the same, $.004 < .007$ determines which is greater)

6) $0.29 > 25\%$ (since $25\% = \frac{25\%}{100\%} = 0.25$ and since $0.29 > 0.25$)

7) $0.5 > 30\%$ (since $30\% = \frac{30\%}{100\%} = 0.3$ and since $0.5 > 0.3$)

8) $1.25 < 210\%$ (since $210\% = \frac{210\%}{100\%} = 2.10$ and since $1.25 < 2.10$)

9) $-0.4 > -0.6$ (since -0.4 is less negative than -0.6, it follows that -0.4 is greater than -0.6)

10) $-0.13 < -0.11$ (since -0.13 is more negative than -0.11, it follows that -0.13 is less than -0.11)

10.3 Comparing Fractions

1) $\frac{1}{2} > \frac{3}{7}$ since $\frac{1}{2} = \frac{1\cdot7}{2\cdot7} = \frac{7}{14}, \frac{3}{7} = \frac{3\cdot2}{7\cdot2} = \frac{6}{14}$, and $\frac{7}{14} > \frac{6}{14}$

2) $\frac{5}{3} > \frac{7}{5}$ since $\frac{5}{3} = \frac{5\cdot5}{3\cdot5} = \frac{25}{15}, \frac{7}{5} = \frac{7\cdot3}{5\cdot3} = \frac{21}{15}$, and $\frac{25}{15} > \frac{21}{15}$

3) $\frac{1}{4} > \frac{1}{5}$ since $\frac{1}{4} = \frac{1\cdot5}{4\cdot5} = \frac{5}{20}, \frac{1}{5} = \frac{1\cdot4}{5\cdot4} = \frac{4}{20}$, and $\frac{5}{20} > \frac{4}{20}$

4) $\frac{5}{6} < \frac{8}{9}$ since $\frac{5}{6} = \frac{5\cdot3}{6\cdot3} = \frac{15}{18}, \frac{8}{9} = \frac{8\cdot2}{9\cdot2} = \frac{16}{18}$, and $\frac{15}{18} < \frac{16}{18}$

5) $-\frac{4}{3} < -\frac{5}{4}$ since $-\frac{4}{3} = -\frac{4\cdot4}{3\cdot4} = -\frac{16}{12}, -\frac{5}{4} = -\frac{5\cdot3}{4\cdot3} = -\frac{15}{12}$, and $-\frac{16}{12} < -\frac{15}{12}$

Note: Since $-\frac{16}{12}$ is more negative than $-\frac{15}{12}$, it follows that $-\frac{4}{3}$ is less than $-\frac{5}{4}$.

Note: If there weren't any minus signs, this inequality would be reversed.

10.4 Comparing Rational Numbers

1) $3.7 > \frac{7}{2}$ since $\frac{7}{2} = \frac{7\cdot5}{2\cdot5} = \frac{35}{10} = 3.5$ and since $3.7 > 3.5$

2) $\frac{9}{20} > 0.42$ since $\frac{9}{20} = \frac{9\cdot5}{20\cdot5} = \frac{45}{100} = 0.45$ and since $0.45 > 0.42$

3) $35\% < \frac{4}{11}$ since $35\% = \frac{35\%}{100\%} = 0.35, \frac{4}{11} = \frac{4\cdot9}{11\cdot9} = \frac{36}{99} = 0.\overline{36}$, and $0.35 < 0.\overline{36}$

4) $3 > \frac{11}{4}$ since $\frac{11}{4} = \frac{11\cdot25}{4\cdot25} = \frac{275}{100} = 2.75$ and since $3 > 2.75$

5) $0.567 < \frac{3}{5}$ since $\frac{3}{5} = \frac{3\cdot2}{5\cdot2} = \frac{6}{10} = 0.6$ and since $0.567 < 0.6$

6) $-1.6 > -\frac{7}{4}$ since $-\frac{7}{4} = -\frac{7\cdot25}{4\cdot25} = -\frac{175}{100} = -1.75$ and since $-1.6 > -1.75$

Note: since -1.6 is not as negative as -1.75, it follows that -1.6 is greater than -1.75

10.5 Comparing Rational and Irrational Numbers

1) $2 > \sqrt{3}$ since $2^2 = (2)(2) = 4$ and since $\sqrt{4} > \sqrt{3}$

2) $2.2 < \sqrt{5}$ since $2.2^2 = (2.2)(2.2) = 4.84$ and since $\sqrt{4.84} < \sqrt{5}$

3) $\sqrt{17} > 4$ since $4^2 = (4)(4) = 16$ and since $\sqrt{17} > \sqrt{16}$

4) $5.4 < \sqrt{30}$ since $5.4^2 = (5.4)(5.4) = 29.16$ and since $\sqrt{29.16} < \sqrt{30}$

5) $\sqrt{79} < 9$ since $9^2 = (9)(9) = 81$ and since $\sqrt{79} < \sqrt{81}$

6) $\sqrt{0.3} > \frac{1}{4}$ since $\frac{1}{4} = \frac{1\cdot25}{4\cdot25} = \frac{25}{100} = 0.25, 0.25^2 = (0.25)(0.25) = 0.0625$ and since $\sqrt{0.3} > \sqrt{0.0625}$

Alternate solution: $\left(\frac{1}{4}\right)^2 = \left(\frac{1}{4}\right)\left(\frac{1}{4}\right) = \frac{1}{16}$ and $0.3 = \frac{3}{10}$, so that $\sqrt{\frac{3}{10}} > \sqrt{\frac{1}{16}}$ since $\frac{3}{10} > \frac{1}{16}$

Note that $\frac{3}{10} > \frac{1}{16}$ because $\frac{3}{10} = \frac{3 \cdot 8}{10 \cdot 8} = \frac{24}{80}$, $\frac{1}{16} = \frac{1 \cdot 5}{16 \cdot 5} = \frac{5}{80}$, and $\frac{24}{80} > \frac{5}{80}$

10.6 Inequalities with Variables

1) $x < 4$ (equivalent to $4 > x$)

Check: $2(3.9) + 7 = 7.8 + 7 = 14.8 < 15$, but $2(4.1) + 7 = 8.2 + 7 = 15.2 > 15$

2) $1 < x$ (equivalent to $x > 1$)

Check: $3(1.1) + 6 = 3.3 + 6 = 9.3$ is less than $9(1.1) = 9.9$,

but $3(0.9) + 6 = 2.7 + 6 = 8.7$ is greater than $9(0.9) = 8.1$

3) $3 > x$ (equivalent to $x < 3$)

Check: 3 is greater than $4(2.9) - 9 = 11.6 - 9 = 2.6$,

but 3 is less than $4(3.1) - 9 = 12.4 - 9 = 3.4$

4) $x > 8$ (equivalent to $8 < x$)

Check: $4(8.1) = 32.4$ is greater than $56 - 3(8.1) = 56 - 24.3 = 31.7$,

but $4(7.9) = 31.6$ is less than $56 - 3(7.9) = 56 - 23.7 = 32.3$

5) $x > 7$ (equivalent to $7 < x$)

Check: $4(7.1) = 28.4$ is greater than $7 + 3(7.1) = 7 + 21.3 = 28.3$,

but $4(6.9) = 27.6$ is less than $7 + 3(6.9) = 7 + 20.7 = 27.7$

6) $x < 5$ (equivalent to $5 > x$)

Check: $6(4.9) - 5 = 29.4 - 5 = 24.4$ is less than 25,

but $6(5.1) - 5 = 30.6 - 5 = 25.6$ is greater than 25

10.7 Inequalities with Variables and Minus Signs

1) $-8 < x$ Check: 8 is greater than $-(-7.9) = 7.9$, but 8 is less than $-(-8.1) = 8.1$

2) $6 > x$ Check: -42 is less than (more negative than) $-7(5.9) = -41.3$,

but -42 is greater than (not as negative as) $-7(6.1) = -42.7$

3) $x > 2$ Check: $1 - 2.1 = -1.1$ is less than (more negative than) -1,

but $1 - 1.9 = -0.9$ is greater than (not as negative as) -1

4) $-8 < x$ Check: 15 is greater than $-9 - 3(-7.9) = -9 - (-23.7) = 14.7$,

but 15 is less than $-9 - 3(-8.1) = -9 - (-24.3) = 15.3$

5) $x > -5$ Check: $-4(-4.9) + 8 = 19.6 + 8 = 27.6$ is less than 28,

but $-4(-5.1) + 8 = 20.4 + 8 = 28.4$ is greater than 28

6) $-1 > x$ Check: -5 is less than (more negative than) $-4(-1.1) - 9 = 4.4 - 9 = -4.6$,

but -5 is greater than (not as negative as) $-4(-0.9) - 9 = 3.6 - 9 = -5.4$

7) $x < -10$ Check: $-3(-10.1) = 30.3$ is greater than $10 - 2(-10.1) = 10 - (-20.2) = 30.2$,

but $-3(-9.9) = 29.7$ is less than $10 - 2(-9.9) = 10 - (-19.8) = 29.8$

8) $x < -6$ Check $5(-6.1) + 10 = -30.5 + 10 = -20.5$ is less than (more negative than) -20,

but $5(-5.9) + 10 = -29.5 + 10 = -19.5$ is greater than (not as negative as) -20

9) $2 < x$ Check: $-18 + 3(2.1) = -18 + 6.3 = -11.7$ is greater than (not as negative as) $-6(2.1) = -12.6$, but $-18 + 3(1.9) = -18 + 5.7 = -12.3$ is less than (more negative than) $-6(1.9) = -11.4$

10) $9 < x$ Check: $-36 - 4(9.1) = -36 - 36.4 = -72.4$ is greater than (not as negative as) $-8(9.1) = -72.8$, but $-36 - 4(8.9) = -36 - 35.6 = -71.6$ is less than (more negative than) $-8(8.9) = -71.2$

11) $x > -3$ Check: $-(-2.9) - 3 = 2.9 - 3 = -0.1$ is greater than (not as negative as) $-9 - 3(-2.9) = -9 - (-8.7) = -9 + 8.7 = -0.3$, but $-(-3.1) - 3 = 3.1 - 3 = 0.1$ is less than $-9 - 3(-3.1) = -9 - (-9.3) = -9 + 9.3 = 0.3$

12) $x < -7$ Check: $-[-(-7.1)] = -(7.1) = -7.1$ is less than (more negative than) -7,

but $-[-(-6.9)] = -(6.9) = -6.9$ is greater than (not as negative as) -7

Note: This problem is much simpler if you realize that $-(-x) = x$, such that the given equation is the same as $x < -7$. It isn't necessary to multiply or divide by negative one.

However, if you wish to multiply or divide by negative one, you must do it twice because the given equation has two minus signs. This reverses the direction of the inequality two times, with the net result of leaving it unchanged.

13) $x > -\frac{10}{3}$ Hints: First cross-multiply to get $-3x < 10$. When you divide by negative three, reverse the direction of the inequality.

14) $\frac{8}{3} < x$ Hints: First cross-multiply to get $-8 > -3x$. When you divide by negative three, reverse the direction of the inequality.

15) $x < 1$ Hints: Subtract $\frac{1}{2}$ from both sides to get $-\frac{x}{3} > -\frac{1}{3}$. When you multiply by negative three, reverse the direction of the inequality. Note: $\frac{1}{6} - \frac{1}{2} = \frac{1}{6} - \frac{3}{6} = -\frac{2}{6} = -\frac{1}{3}$.

16) $4 > x$ Hints: Subtract $\frac{1}{3}$ from both sides to get $-\frac{1}{x} < -\frac{1}{4}$. Cross-multiply to get $-4 < -x$. When you multiply (or divide) by negative one, reverse the direction of the inequality.

GLOSSARY

absolute value: the value of a number without its sign, indicated by vertical lines surrounding a number. For example, $|-7| = 7$.

additive inverse: adding a negative value (or subtracting a positive value) is the opposite of adding a positive value in the sense that $x + (-x) = x - x = 0$.

associative property: the result of adding or multiplying three (or more) numbers does not depend on how the numbers are grouped: $(x + y) + z = x + (y + z)$ and $(xy)z = x(yz)$.

base: a number that has an exponent (or power). For example, in 5^4 the base is 5.

brackets: the symbols [and], used when an expression would otherwise consist of inside and outside parentheses. An example is $[x^3 + (y + z)^3]^2$.

coefficient: a number that multiplies a variable. For example, in $8x$ the coefficient is 8.

colon: the symbol : that appears between two numbers in a ratio. For example, 3:4 represents the ratio of three to four.

common denominator: when two fractions have the same denominator, like $\frac{3}{8}$ and $\frac{6}{8}$.

commutative property: the result of adding or multiplying two numbers does not depend on the order: $x + y = y + x$ and $xy = yx$.

constant: a fixed value. For example, in $x + 4$ the number 4 is a constant.

conversion: the process of changing the form of a quantity, such as converting a decimal to a percent or converting feet to inches.

cross multiply: multiply the numerator of the left fraction with the denominator of the right fraction, then multiply the denominator of the left fraction with the numerator of the right fraction, and set these quantities equal. For example, $\frac{3}{x} = \frac{2}{5}$ becomes $3(5) = 2x$.

cube: raise a number to the power (or exponent) of three. For example, 6^3 is 6 cubed.

cube root: determine which number cubed equals a specified value. For example, $\sqrt[3]{8}$ is the cube root of 8. Note that $\sqrt[3]{8} = 2$ because $2^3 = 8$.

decimal: a fraction where the denominator is a power of ten. For example, $0.73 = \frac{73}{100}$.

decimal point: a low dot . that appears between the whole part and the fractional part. For example, 3.9 is the same as $3 + .9$ (or $3 + 0.9$) and as $3\frac{9}{10}$.

decimal position: the location of the decimal point. For example, 1.375 and 137.5 differ by a factor of 100 because of the difference in their decimal positions.

denominator: the number at the bottom of a fraction. For example, in $\frac{2}{3}$ the denominator is 3.

difference of squares: the formula $x^2 - y^2 = (x + y)(x - y)$.

dime: a coin in the United States with a value equal to ten cents.

discount: a reduction in price often stated as a percentage. For example, a store might offer a 25% discount on selected merchandise.

distance: the length between two points, or how far an object has traveled.

distributive property: $x(y + z) = xy + xz$.

division symbol: the $\div$ or / symbol that appears between numbers, like $18 \div 6$ or 18/3. If a variable is involved, it is usually expressed as a fraction like $\frac{x}{2}$ (instead of $x \div 2$).

dollar: United States currency with a value of 100 cents.

elapsed time: the amount of time that has passed, such as the duration of time for which an object has been moving.

equal sign: the $=$ symbol that represents the equality between two expressions or numbers.

equation: a mathematical statement that sets two expressions or numbers equal. An equation always contains an equal sign, such as $2x - 3 = 5$.

equivalent fractions: two fractions that equal the same value, such as $\frac{3}{4}$ and $\frac{9}{12}$. Note that these fractions are equal because $\frac{3}{4} = \frac{3 \times 3}{4 \times 3} = \frac{9}{12}$.

even: a number that is evenly divisible by 2, such as 2, 4, 6, 8, 10, 12, 14, 16, etc.

exponent: the number of times the base appears multiplied together. The exponent is indicated by a small number to the top right of the base. For example, in 5^4 the exponent is 4 and means $5 \times 5 \times 5 \times 5$. Another word for exponent is power.

expression: a mathematical idea involving constants, variables, and operators, but which does not have an equal sign or inequal sign (like $<$ or $>$). An example is $3x^2 - 7$.

factor: a number being multiplied like the 5 and 6 in $5 \times 6 = 30$.

factor an expression: apply the distributive property in reverse, like $3x^2 - 6x = 3x(x - 2)$.

factor a perfect square: pull a perfect square out of a radical, like $\sqrt{12} = \sqrt{4}\sqrt{3} = 2\sqrt{3}$.

factorization: numbers multiplied together to make another number. For example, the prime factorization of 18 is written as $2 \times 3 \times 3$.

FOIL: an abbreviation for "first, outside, inside, last" which is used to help students remember that $(w + x)(y + z) = wy + wz + xy + xz$.

fraction: a number of the form $\frac{2}{3}$ or 2/3.

greater than sign: the symbol $>$ indicating that the left value is larger than the right value, as in $7 > 4$.

greatest common factor (GCF): the largest factor that is common to two different integers. For example, 5 is the GCF of 15 and 20 since $15 = 3 \times 5$ and $20 = 4 \times 5$.

hundredths place: the second digit after a decimal point. For example, in 1.234 the 3 is in the hundredths place.

identity property: adding zero or multiplying by one have no effect: $x + 0 = x$ and $1x = x$.

improper fraction: a fraction for which the value of the numerator exceeds the value of the denominator, such as $\frac{9}{4}$.

inequality: a mathematical statement where the two sides aren't equal (like $x \neq 3$) or where one side is less than or greater than the other (like $x > 2$).

integer: numbers like 0, 1, 2, 3, etc. and their negatives ($-1, -2, -3$, etc.).

interest: money that is earned from savings or investments, or that is paid to take out a loan.

inverse property: the opposite of an operation. For example, adding a negative value is the inverse of addition because $x + (-x) = 0$, and multiplying by the reciprocal of a value is the inverse of multiplication because $x\left(\frac{1}{x}\right) = \frac{x}{x} = 1$.

investment: money that is put into a savings account, business, stocks, etc. with the potential to make a profit (but sometimes also at risk of suffering a loss).

irrational number: a number that can't be expressed in the form of an integer or as the ratio of two integers, such as $\sqrt{2}$ (but not like $\sqrt{4}$ since $\sqrt{4} = 2$).

isolate the unknown: apply operations to both sides of an equation and combine like terms in order to get the variable all by itself on one side of the equation.

leading zero: a zero that comes after a decimal point and before nonzero digits. For example, 0.0008743 has 3 leading zeroes. (The 0 *before* the decimal point doesn't count.)

least common multiple (LCM): the smallest integer that is a multiple of two other integers. For example, the LCM of 6 and 8 is 24 since $6 \times 4 = 24$ and $8 \times 3 = 24$.

less than sign: the symbol $<$ indicating that the left value is smaller than the right value, as in $3 < 8$.

like terms: terms in an expression, equation, or inequality where the same variable is raised to the same power, such as the two terms in $7x^2 + 5x^2$ (since both involve x^2).

linear proportion: a relationship between two quantities where an increase in one quantity results in an increase in the other quantity by the same factor (and similarly for a decrease).

lowest common denominator (LCD): the smallest common denominator that two fractions can form. For example, the LCD of $\frac{5}{6}$ and $\frac{4}{9}$ is 18 since $\frac{5}{6} = \frac{5 \times 3}{6 \times 3} = \frac{15}{18}$ and $\frac{4}{9} = \frac{4 \times 2}{9 \times 2} = \frac{8}{18}$.

middle dot: the symbol · sometimes used to represent multiplication between numbers, like 7·8 = 56 (not to be confused with a decimal point . which is a low dot, like 3.14).

mixed number: a number that includes an integer plus a fraction, such as $7\frac{3}{4}$ (which equals seven plus three-fourths).

moving the decimal point: multiplying by a power of 10 has the effect of moving the decimal point to the right, while dividing by a power of 10 has the effect of moving the decimal point to the left. For example, $8.24715 \times 10^3 = 8.24715 \times 1000 = 8247.15$.

multiplication symbol: the times symbol ($\times$) is seldom used in the context of algebra because it could easily be confused with the variable x. Instead, multiplication between numbers is represented with a middle dot, like 3·4 = 12, or parentheses, like 3(4) = 12 or (3)(4) = 12. When multiplying variables, no symbol is used, such as $3xy$.

multiplicative inverse: multiplying by the reciprocal of a value (or dividing by the value) is the opposite of multiplication in the sense that $x\left(\frac{1}{x}\right) = \frac{x}{x} = x \div x = 1$.

nickel: a coin in the United States with a value equal to five cents.

numerator: the number at the top of a fraction. For example, in $\frac{2}{3}$ the numerator is 2.

odd: a number that isn't evenly divisible by 2, such as 1, 3, 5, 7, 9, 11, 13, 15, etc.

operation: a mathematical process such as addition, subtraction, multiplication, or division.

order of operations: the order in which arithmetic operations are carried out (parentheses first, then exponents, then multiplication and division from left to right, and then addition and subtraction from left to right).

parentheses: the symbols (and) placed around an expression, like $3x(x^2 - 5)$. The singular form of the word is parenthesis, whereas the plural form is parentheses.

part to part: a ratio between two parts, such as the ratio of girls to boys (since girls and boys are both parts of the total population).

part to whole: a ratio between one part and the whole, such as the ratio of girls to the total number of students.

PEMDAS: an abbreviation for "parentheses, exponents, multiply/divide from left to right, and add/subtract from left to right" used to help students remember the order of operations.

penny: a coin in the United States with a value equal to one cent.

percent: a specified fraction of 100. For example, 37% means 37 out of 100.

percentage: an unspecified amount, like "a percentage of the students."

perfect square: an integer that equals the square of another integer. For example, 9 is a perfect square because $3^2 = 3 \times 3 = 9$.

place value: the position of a digit in a number. For example, in 12.345, the 1 is in the tens place, the 2 is in the units place, the 3 is in the tenths place, the 4 is in the hundredths place, and the 5 is in the thousandths place.

power: the number of times the base appears multiplied together. The power is indicated by a small number to the top right of the base. For example, in 5^4 the power is 4 and means $5 \times 5 \times 5 \times 5$. Another word for power is exponent.

prime factorization: prime numbers multiplied together to make another number. For example, the prime factorization of 18 is written as $2 \times 3 \times 3$.

prime number: a whole number that is evenly divisible only by itself and one, such as 2, 3, 5, 7, 11, 13, 17, and 19. (The in-between numbers, such as 4, 6, 8, 9, and 10, are called composite numbers).

principal: the amount of money invested or borrowed before the interest rate is applied.

proper fraction: a fraction for which the value of the numerator is smaller than the value of the denominator, such as $\frac{4}{9}$.

proportion: an equality between two ratios or rates.

quarter: a coin in the United States with a value equal to twenty-five cents.

radical: the $\sqrt{}$ symbol used to indicate a square root or other root (such as $\sqrt[3]{}$).

rate: a fraction made by dividing two quantities that have different units, like $\frac{100 \text{ miles}}{3 \text{ hours}}$.

ratio: a fixed relationship expressed in the form $x{:}y$. For example, the ratio of fingers to hands is 10:2 (which reduces to 5:1) for a typical human being.

rational number: a number that can be expressed in the form of an integer or as the ratio of two integers, such as 5 or $\frac{7}{3}$ (but not $\sqrt{2}$).

rationalize the denominator: make the denominator rational, like $\frac{1}{\sqrt{3}} = \frac{1}{\sqrt{3}} \times \frac{\sqrt{3}}{\sqrt{3}} = \frac{\sqrt{3}}{3}$.

real number: a number that is rational or irrational, and which does not have an imaginary part. For example, 3, $\sqrt{5}$, and $1 + \sqrt{2}$ are real numbers (whereas $\sqrt{-1}$ is imaginary because no real number multiplied by itself can be negative).

reciprocal: one divided by a number. For a fraction, this means to swap the numerator and denominator. For example, $\frac{3}{2}$ is the reciprocal of $\frac{2}{3}$, and $\frac{1}{4}$ is the reciprocal of 4.

reduced fraction: the simplest form of a fraction, in which the numerator and denominator do not share a common factor. For example, $\frac{1}{3}$ is the reduced form of $\frac{4}{12}$ (since $\frac{4}{12} = \frac{4 \div 4}{12 \div 4} = \frac{1}{3}$).

remainder: the amount left over when one number is divided by another number. For example, $23 \div 5$ equals 4 with a remainder of 3 because $5 \times 4 = 20$ and $23 - 20 = 3$.

repeating decimal: a digit or a group of digits that repeat forever in a decimal number, such as $0.\overline{36} = 0.3636363636...$

root: the opposite of a power. A general root asks, "Which number raised to the indicated power equals the value under the radical?" For example, $\sqrt[3]{8} = 2$ because $2^3 = 8$.

sales tax: an amount that the government collects in addition to the sales price, usually as a percent of the subtotal. For example, a sales tax of 8% means to add 8% to the subtotal.

scientific notation: using a power of ten in order to position a decimal point immediately after the first digit of a number. For example, the number 72,000 can be expressed with scientific notation as 7.2×10^4 (since $7.2 \times 10^4 = 7.2 \times 10,000 = 72,000$).

simple interest: money that is earned every period (such as a year, for an annual interest rate) based on a percentage of the principal. For example, if $200 is invested in a savings account that earns simple interest at a rate of 3%, in one year the interest earned is $6 (which is 3% of $200, since $0.03 \times \$200 = \$6$).

simplify: make an expression simpler. For example, $5x - 2 + 3x$ simplifies to $8x - 2$.

slash: the / symbol sometimes used to indicate division or a fraction like 3/5.

solve: determine the value(s) of the variable(s) by following a procedure (like isolating the unknown).

speed: a measure of how fast an object moves, like 40 mph (miles per hour).

square: raise a number to the power of two. For example, 4^2 is 4 squared (meaning 4×4).

square root: a number that when multiplied by itself makes the value under the radical. For example, $\sqrt{25} = \pm 5$ because $5^2 = 5 \times 5 = 25$ and $(-5)^2 = (-5) \times (-5) = 25$.

subtotal: an amount prior to the total, such as the amount before sales tax is added.

tax: money collected by the government, such as sales tax, income tax, or property tax.

tenths place: the digit after a decimal point. For example, in 1.23 the 2 is in the tenths place.

term: an expression separated from other expressions by plus ($+$), minus ($-$), equal ($=$), or inequal ($<$ or $>$) signs in an algebraic statement. For example, $4x + 7 = 5x$ consists of three terms ($4x$, 7, and $5x$).

thousandths place: the third digit after a decimal point. For example, in 1.234 the 4 is in the thousandths place.

trailing zero: a zero that comes at the end of a number (and after a decimal point). For example, 0.0000007500 has 2 trailing zeroes (which come after the 5).

unit: a standard value for measurement, such as a meter, foot, or a second.

unit ratio: an equivalent ratio where one of the numbers (typically, the second number) is equal to one. For example, 1.5:1 is the unit ratio of 3:2 (found by dividing each number by 2).

units place: the digit before the decimal point. For example, in 1.2, the 1 is in the units place.

unity: the number one.

unknown: a quantity like x which needs to be solved for.

unlike terms: terms with different powers of the variable, such as $3x^2$ and $3x$.

variable: an unknown quantity represented by a letter like x or y.

whole number: a number that is whole like 1, 2, 3, etc.

whole to part: a ratio between the whole and one part, such as the ratio of the total number of students to the number of girls.

INDEX

WAS THIS BOOK HELPFUL?

A great deal of effort and thought was put into this book, such as:

- Breaking down the solutions to help make the math easier to understand.
- Careful selection of examples and problems for their instructional value.
- Full solutions to the exercises included in the answer key.

If you appreciate the effort that went into making this book possible, there is a simple way that you could show it:

Please take a moment to post an honest review.

For example, you can review this book at Amazon.com or Barnes & Noble's website at BN.com.

Even a short review can be helpful and will be much appreciated. If you're not sure what to write, following are a few ideas, though it's best to describe what's important to you.

- How much did you learn from reading and using this workbook?
- Were the solutions at the back of the book helpful?
- Were you able to understand the solutions?
- Was it helpful to follow the examples while solving the problems?
- Would you recommend this book to others? If so, why?

Do you believe that you found a mistake? Please email the author, Chris McMullen, at greekphysics@yahoo.com to ask about it. One of two things will happen:

- You might discover that it wasn't a mistake after all and learn why.
- You might be right, in which case the author will be grateful and future readers will benefit from the correction. Everyone is human.

ABOUT THE AUTHOR

Dr. Chris McMullen has over 20 years of experience teaching university physics in California, Oklahoma, Pennsylvania, and Louisiana. Dr. McMullen is also an author of math and science workbooks. Whether in the classroom or as a writer, Dr. McMullen loves sharing knowledge and the art of motivating and engaging students.

The author earned his Ph.D. in phenomenological high-energy physics (particle physics) from Oklahoma State University in 2002. Originally from California, Chris McMullen earned his Master's degree from California State University, Northridge, where his thesis was in the field of electron spin resonance.

As a physics teacher, Dr. McMullen observed that many students lack fluency in fundamental math skills. In an effort to help students of all ages and levels master basic math skills, he published a series of math workbooks on arithmetic, fractions, long division, algebra, geometry, trigonometry, and calculus entitled *Improve Your Math Fluency*. Dr. McMullen has also published a variety of science books, including astronomy, chemistry, and physics workbooks.

Author, Chris McMullen, Ph.D.

PUZZLES

The author of this book, Chris McMullen, enjoys solving puzzles. His favorite puzzle is Kakuro (kind of like a cross between crossword puzzles and Sudoku). He once taught a three-week summer course on puzzles. If you enjoy mathematical pattern puzzles, you might appreciate:

300+ Mathematical Pattern Puzzles

Number Pattern Recognition & Reasoning

- Pattern recognition
- Visual discrimination
- Analytical skills
- Logic and reasoning
- Analogies
- Mathematics

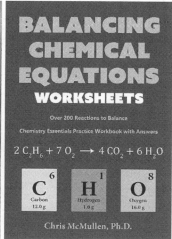

ARITHMETIC

For students who could benefit from additional arithmetic practice:

- Addition, subtraction, multiplication, and division facts
- Multi-digit addition and subtraction
- Addition and subtraction applied to clocks
- Multiplication with 10-20
- Multi-digit multiplication
- Long division with remainders
- Fractions
- Mixed fractions
- Decimals
- Fractions, decimals, and percentages
- Grade 6 math workbook

www.improveyourmathfluency.com

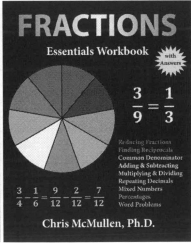

MATH

This series of math workbooks is geared toward practicing essential math skills:

- Prealgebra
- Algebra
- Geometry
- Trigonometry
- Calculus
- Fractions, decimals, and percentages
- Long division
- Multiplication and division
- Addition and subtraction
- Roman numerals

www.improveyourmathfluency.com

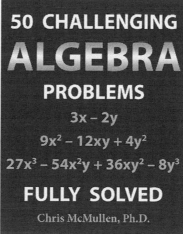
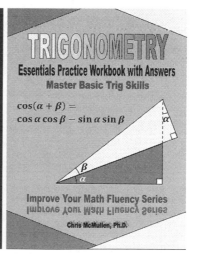

THE FOURTH DIMENSION

Are you curious about a possible fourth dimension of space?

- Explore the world of hypercubes and hyperspheres.
- Imagine living in a two-dimensional world.
- Try to understand the fourth dimension by analogy.
- Several illustrations help to try to visualize a fourth dimension of space.
- Investigate hypercube patterns.
- What would it be like to be a 4D being living in a 4D world?
- Learn about the physics of a possible four-dimensional universe.

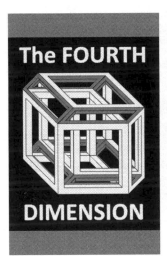

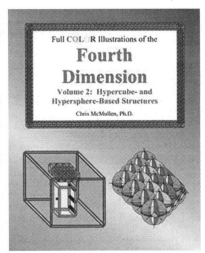

SCIENCE

Dr. McMullen has published a variety of **science** books, including:

- Basic astronomy concepts
- Basic chemistry concepts
- Balancing chemical reactions
- Calculus-based physics textbooks
- Calculus-based physics workbooks
- Calculus-based physics examples
- Trig-based physics workbooks
- Trig-based physics examples
- Creative physics problems
- Modern physics

www.monkeyphysicsblog.wordpress.com

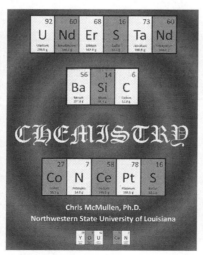
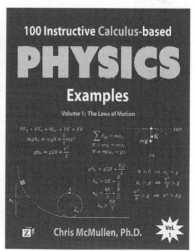

Made in the USA
Monee, IL
11 March 2024

54809271R00125